Preliminary Edition

Experiences in Statistics

Dennis C. Gilliland
Michigan State University

KENDALL/HUNT PUBLISHING COMPANY
2460 Kerper Boulevard P.O. Box 539 Dubuque, Iowa 52004-0539

Contents

Introduction

This book was written to serve as a short, companion text for standard introductory texts typically used in basic one or two term courses in statistics at the pre-calculus level. Some standard texts neglect or quickly pass over certain issues that are fundamental to the proper application of statistics and choose rather to emphasize the formal inferential structure of estimation and hypothesis testing. The examples in some of these standard texts are sometimes superficial and unrealistic examples that serve only to illustrate the manipulative aspects of the numerous formulas that fill the pages. When case studies are included, they are sometimes cold and complicated choices from the literature with which the authors have no intimate knowledge.

The readings in this companion text are based on the author's experiences in applying probability and statistics in a variety of situations. Many of these experiences involved matters that were of concern to courts or involved administrative hearings.

In these arenas, the applications are expected to be critically analyzed. Therefore, concern for the fundamental issues that underpin formal inference is essential. Words and concepts and theories must be questioned. Sometimes the issues get deep and to the very meaning of inductive reasoning. I will not go that far in describing any of the experiences. However, I believe that the experiences and the exposition will show the reader a lively and exciting side of statistics.

In this book, I have used standard notations whenever symbols are used in describing the experiences. I have not attempted to develop ideas in a sequential manner as is the custom of mathematical and linear developments of a set of ideas and theories. Rather, I discuss some ideas before they are completely defined. At these points, unless the reader has worked with another text, he/she may acquire only an intuitive notion of what the author is discussing. (Henry Ford once said "We go forward without facts, and we learn facts as we go along." In some places in this book, we go forward without definitions, and we develop concepts as we go along.)

One must realize that written language is necessarily circular. Dr. Robert Carson, a most insightful and intelligent person, told me of an "abridged" dictionary where "gorse" is defined as "furze" and "furze" is defined as "gorse". The words I use will have no meaning to you unless we share some experiences and feelings outside of language. With this book and its informal style, I emphasize experience over the formal manipulative use of language. After you are finished reading this book, you should have a feel for probability and statistics, their use, and their importance. Hopefully, some of you will go on to study the theory of probability and statistics and will better appreciate that development because of the experiences that we are sharing here.

The exposition surrounding the experiences, examples and ideas in this book is reasonably self-contained so that most of the readings can stand on their own. In fact, some teachers may choose to assign all of this book before starting study from a standard text. I believe that the readings can be used effectively in this way as well as in parallel with study from a standard text. The readings should serve to illustrate the excitement and urgency that sometimes surround the application of probability and statistics and should reinforce many of the concepts that authors of standard texts are trying to get across.

I will present the experiences in an informal conversational style. I find it easiest to write in this style and believe that my choice to do so will lighten-up what is commonly thought of as a dry subject. I have lots of evidence that there is a "widely-held" belief that "statistics" is a dry subject. This evidence comes from my personal observation and the collective anecdotal experience of the faculty and other professionals that I have met. It seems that admitting that one is a "statistician" at a social affair is tantamount to inviting a cessation of conversation, usually after a person has reacted by indicating having had an earlier experience with the subject in college. Apparently, the word "statistics" is associated with very negative experiences that the person wishes not to recall. Then there are the countless jokes about the subject. A person after having been informed by his doctor of having but an afternoon to live, indicated that he was going to spend that afternoon attending a lecture on statistics. When asked why, he replied "a statistics lecture can make an hour seem like two weeks."

The book is motivated in part by the author's interest in the current and extremely important movement in America to improve quality and productivity. In particular, the author is interested in the contributions of Dr. W. Edwards Deming to this movement. I have found that my students in an introductory course in statistics for business majors appreciate learning about Dr. Deming and the quality movement in America. This book will provide these students with relevant readings and some examples of applications of statistics in the quality movement.

I believe that one must understand certain fundamental ideas in statistics in order to understand Dr. Deming's management theories. I hope that this book will help in providing some insights into those fundamental ideas and will help readers more fully appreciate the subtleties in Scherkenbach (1988) and Walton (1986).

There are books that discuss very meaningful and exciting experiences in the application of statistics. Rao (1989) is filled with wonderful examples that will stretch the reader's appreciation of the care that must accompany the application of statistics. Examples will take you to questions of the very meaning of inference. At the elementary level, each of the books by Folks (1981), Moore (1986) and Freedman, Pisani and Purvis (1979) can easily stand alone in courses as a source of basic theory and examples of the application of methods. Each is a much more ambitious and comprehensive treatment of statistical experiences than what I attempt in this small tome. Wallis and Roberts (1956) is an earlier development of basic statistics that integrated meaningful examples with the development of a body of theory and methods. The readings compiled by Tanur et. al. (1989) are very well-written descriptions of statistical studies, some of which had impact at the national and world level. Gastwirth (1988) has woven an introduction to theory and methods of statistics with interesting applications to health, law, evidence and public policy.

What I attempt is small in comparison; but, hopefully, this book will provide interesting reading of topical material for the student trying to learn more about the application of probability and statistics. I hope that the reader will not find it presumptuous of the author to write at length about these experiences. Of course, all opinions expressed are those of the author.

Chapter 1 is included to give the reader my brief summary of the background of Dr. Deming and the quality movement.

Chapter 2 gets into population, frame, random sampling, and statistical estimation as common concepts and tools in enumerative studies. Inference based on random samples is discussed within the context of certain experiences. Sampling error and measurement error are discussed and illustrated.

Chapter 3 is a bit of a digression to convey my thoughts on statistics and the law. In some sense, this chapter is a call for professional standards and a code of ethics for statisticians.

Chapter 4 introduces some elementary ideas of probability and four important families of probability models. These models are used in a variety of settings as a tools for modeling variability.

Chapter 5 discusses some tools and ideas that are important to the quality movement. An introduction to the philosophy of continuous improvement is given as well as insights into statistical process control, capability analysis and error transmission.

Chapter 6 concludes the readings with a brief discussion of some of the central themes that emerge from all of the experiences. For examples, the fact that data and results must have proper pedigrees in order to be useful and the fact that there must be proper documentation of these pedigrees for the users of the data and results, that is, the scientists, managers, workers, judges, etc. who are the consumers of statistics. I try to get across the importance of advance agreements and cooperation among adversaries in regard to procedures to be used in decision-making proceedings. If the readers of this book come away from it with a sense and understanding of the importance of pedigree and cooperation, I will feel that the book has made a contribution.

We will number sections within a chapter in the form m.n, where the root m refers to the chapter and the extension n refers to the section. Examples, exercises and figures will be numbered in the same format. Articles and books that are cited are listed by chapter.

Finally, I wish to thank my colleagues for useful discussions over the years and for having brought to my attention some of the newspaper articles that I cite in this book.

References

Folks, J. Leroy, *Ideas of Statistics*, Wiley, New York, 1981.

Freedman, David, Pisani, Robert and Purves, Roger, *Statistics*, W. W. Norton, New York, 1978.

Gastwirth, Joseph L., *Statistical Reasoning in Law and Public Policy, Vols. 1 and 2*, Academic Press, Boston, 1988.

Moore, David S., *Statistics: Concepts and Controversies, Second Edition*, W. H. Freeman, San Francisco, 1986.

Rao, C. Radhakrishna, *Statistics and Truth: Putting Chance to Work*, Council of Scientific & Industrial Research, New Delhi, India, 1989.

Scherkenbach, William W., *The Deming Route to Quality and Productivity; Road Maps and Roadblocks*, CEEPress Books, George Washington University, Washington, D.C., 1988.

Tanur, Judith M., Mosteller, Frederick, Kruskal, William H., Lehmann, Erich L., Link, Richard F., Pieters, Richard S. and Rising, Gerald R., *Statistics: A Guide to the Unknown, Third Edition*, Wadsworth & Brooks/Cole, Pacific Grove, CA, 1989.

Wallis, W. Allen and Roberts, Harry V., *Statistics: A New Approach*, The Free Press, New York, 1956.

Walton, Mary, *The Deming Management Method*, Dodd, Mead & Company, New York, 1986.

Dr. W. Edwards Deming —
The Man and His Mission

This chapter is made-up of the revised text of a talk that I gave on Dr. W. Edwards Deming. This man and his thinking have greatly influenced the way that statistics is applied. The references provided at the end of the chapter will give you entries into the literature where much more can be learned about Dr. Deming.

Section 1.1 Introduction

I thank you for the opportunity to share some of my thinking and knowledge in regard to the management ideas of Dr. W. Edwards Deming. Dean Hoppensteadt saw an article in the N.Y. Times (March 7, 1989) which discussed Dr. Deming and the University of Tennessee and suggested that I come today. Later I will distribute a copy of that article, one from Business Month of October, 1988 on Dr. Deming, and a copy of Dr. Deming's Fourteen Points.

My talk has these sections:

- Dr. Deming's background through 1946.
- The concept of statistical process control (SPC) which seems to have greatly influenced Dr. Deming's thinking and approach to management.
- The media discovery of Dr. Deming in 1981, and what he had accomplished in post-war Japan.
- Dr. Deming's management advice 1981+ . . . what he is doing now.

I will move at a rapid pace to try to capture in 20 minutes the flavor of the man who dispenses his ideas in 4-day seminars about the country.

Section 1.2 Deming's Background

Dr. Deming was born in the year 1900 and was raised on the virtual frontier in Wyoming. He attended the University of Wyoming and later earned the Ph.D. in mathematical physics at Yale. Between degrees he had worked for Western Electric in Chicago and there was encouraged to go further with his formal education. After the Ph.D., Dr. Deming went to work in the federal government, with positions in the USDA and later the Bureau of the Census. He retired from government work about 1946 to begin a career in consulting.

Dr. Deming has held various positions at universities and has published many papers reflecting his ideas on the theory and proper application of statistical methods, on ethics and professional standards, on the statistical profession, and on sampling theory. He has earned the respect of both the academic community in statistics and those who are most concerned with the application of the subject.

Section 1.3 Deming and Statistical Process Control

During the 1920's and 1930's at Bell Laboratories, Shewhart was developing the concepts of using control charts to study industrial processes. Dr. Deming traveled regularly from Washington, D.C. to New York to meet with Shewhart and learn about his theory. It is clear that Dr. Deming's current thinking in regard to processes and their management is greatly influenced by this early contact. We digress a bit to learn about statistical process control (SPC).

Daniel Boorstin gives his interpretation of the impact of SPC on American history and technology in Chapter 22 of his 1973 book, "The Americans: The Democratic Experience". That chapter is entitled "Making Things No Better Than They Need To Be". The chapter contrasts the making of things uniform within limits (one goal of SPC) to making things unique, the goal of the artisan or craftsman. Obviously, for mass production of units and multiple part assemblies, production of many units uniform within limits is desirable. Naturally, SPC was thought of as just one other tool to increase output and decrease costs. One can see that this concept of SPC is consistent with the spirit of the teachings of Frederick Winslow Taylor, the first authority on scientific industrial management. (At the beginning of this century, Taylor's teachings became fashionable with their emphasis on reducing work to mechanized tasks through time study.) The potential of SPC as a tool for management and for increasing quality was not fully realized.

There were early voices speaking out on the quality issue. Although Boorstin interprets SPC as having the objective of making things no better than they need to be, he writes that George Radford, an American engineer, in his 1922 paper "The Control of Quality in Manufacturing" states that there was a disproportionate emphasis on quantity of output and argues that "increased output and decreased costs are more certainly attained when manufacturing problems are approached with quality, instead of quantity, as the primary guide and objective."

SPC is a method to help control processes. It uses simple control charts. Data are gathered on critical characteristics of the parts being produced and plotted in time sequence. Upper and lower control limits are determined as limits of the characteristics reflecting the inherent variability of the process. An "in-control" or "stable" state for the process is defined operationally through a series of rules for examination of the pattern of plotted points.

If you wish, think of a distribution of a key characteristic of the parts being produced when the process is stable. Then as long as the process remains in-control, the characteristics of the units being produced will act like a random sample from the distribution; essentially all will fall within certain control limits, and they will not exhibit special patterns. An "out-of-control" situation is signaled by departure from the stable pattern at a given time. When the process signals out-of-control, SPC calls for searching for a special cause for the extreme variation or trend and the elimination of that cause. Statistical analysis of data, problem solving using a team of operators, engineers and statisticians, theories of cause and effect, and designed experiments are fundamental tools for determining special causes of variation. The manager, who is responsible for the process, should be concerned with improving the quality of his/her output by taking fundamental actions on the process when it is in-control . . . or finding a process that is inherently better. The operator and manager should be concerned with keeping the process in-control.

The packagers of words now prefer "quality improvement" to "quality control" to reflect the positive rather than the negative. Quality drives costs down by eliminating the need for mass inspection and sorting parts.

Dr. Deming takes the ideas of SPC to other levels of application. A process may be the hiring of employees, the teaching of a class, the ordering of textbooks, the arranging of class schedules—all processes within larger processes that are part of an organization's activities. Dr. Deming believes that the lack of quality we see about us is @85% the responsibility of the managers and @15% the responsibility of the workers within the process.

A common error in management that Dr. Deming often sees is illustrated by his famous bead experiment. Walton (1986, Chapter 4) describes the experiment in great detail and refers to it as the "parable of the beads." I had a chance to witness this experiment in July 1988 at the General Motors Technical Center in Warren, Michigan. Dr. Deming calls for volunteers from his audience and assigns some to be willing workers, two to be inspectors, and another to announce the results of the inspections. The task that is assigned to the willing workers is the selection of beads from a plexiglass container that has white and red beads in the ratio 4 to 1. Dr. Deming gives very careful instructions on how the beads are to be selected. A paddle with indentations is to be passed through the beads and lifted out at a particular angle. The result is the selection of 50 beads from those in the plexiglass container. The red beads represent defective (nonconforming product) and the white beads represent good (conforming) product. Each worker receives the same instructions and is well-motivated to produce good parts. After willing worker Mike performs his task, the inspector counts the number of defectives and the tally is announced, say "12". Then it is Joan's turn, then Jim's and finally Bill's. Their results are "7", "11" and "14". Dr. Deming as the production manager praises Joan for her good work and warns Bill that unless things improve, he will be given time off without pay. In the next cycle, there is the variation that you might expect. Perhaps a different person is praised and a different person is criticized. Dr. Deming takes this through several cycles. The workers were not contributing anything but their best efforts at producing good product. However, the system they were placed in did not allow them to improve or to influence to any measurable extent the outcomes for which they were variously praised and criticized.

The point is that here management did not understand that the variation that was seen was part of the system it had created and for which it was responsible. Management did not understand the system and its natural variation when in-control, and it held the workers responsible for the variation.

After the bead experiment and before the next session at the Deming Seminar, the data from the experiment are plotted versus the time order in which the cycles occurred. Control limits are calculated and, as expected, the variation is seen to be common cause variation; there are no special causes of variation apparent. There was no blame or praise to be given to the willing workers. Rather, the production of defectives (red beads) was solely the responsibility of the managers who were responsible for the production system.

Section 1.4   Discovery of Deming; Deming in Post-War Japan

Dr. Deming is now given a lot of respect in America. He is widely recognized as a statistical and management guru. He was not so widely known prior to 1981. In that year, NBC interviewed Dr. Deming on national television. The nation's automobile industry was coming to grips with the fact that Japanese competition was formidable; cars made in Japan were gaining a larger and larger market share even as the energy crunch was phasing out. NBC had been told about this 81 year old gruff and opinionated American statistician in Washington, D.C. who, supposedly, had something to do with the Japanese success. During the interview, Dr. Deming spoke frankly about what was wrong with the American automotive industry and its management. Following the interview, his phone was ringing "off the wall" with American companies wishing to solicit his advice and consultation.

What was right in Japan that made it receptive to Dr. Deming and his ideas? What was wrong in America that had caused it to essentially ignore what he had to say? These questions are topics for much thought and research; much has been written in answer to these questions. We have not the time to do much more than to note a few facts concerning Dr. Deming and Japan. Dr. Deming had gone to Japan to work on the census and while there he was invited to consult on the use of statistics in quality and process control. The Japanese engineers, statisticians and managers had a tremendous interest in the ideas of Dr. Deming and saw the advantages to be gained through a coordinated effort at producing quality product the first time rather than by sorting out inferior product. At the same time, Japanese scientists such as Dr. Kaoru Ishikawa were promoting ideas and systems that would allow the results of statistical thinking and planned experiments to play the lead in on-line and off-line quality control efforts. The visibility of statistics was greatly increased in Japan, and Dr. Deming was there to provide clear thinking on its proper application. David Halberstam in Chapter 17 of his book "The Reckoning" details some of the impact of Dr. Deming in Japan and states "With the possible exception of Douglas MacArthur he was the most famous and most revered American in Japan during the postwar years." Beginning in 1951, the Japanese have annually awarded The Deming Prize. This is a medal named in his honor and given to those companies that attain the highest level of quality. In America, the Malcolm Baldridge Award has been recently created to recognize similar achievement by American companies.

Section 1.5   Deming's Management Advice

After the NBC interview, many American companies hired Dr. Deming and heard his words. Because of his harsh opinion of American management, what he had to say was a bitter pill for some. However, many converts have been won and there is now the appearance of real efforts at creating management systems that are based on the theories and humane ideas of Dr. Deming. In these systems, statistical thinking plays a vital role.

Please read Dr. Deming's book "Out of the Crisis". William Scherkenbach, his former NYU student, has written a book entitled "The Deming Route to Quality and Productivity". He discusses and exemplifies the famous Fourteen Points", one-at-a-time. Dr. Deming's forceful insistence on his Fourteen Points reminds me of a quote from Clemenseau. Woodrow Wilson was insisting that any settlement of World War I would have to be based on his Fourteen Points. Clemenseau saw this as an obstacle to the settlement eventually won at Versailles and, in frustration, remarked something to the effect that "God had only Ten Points".

Dr. Deming's Fourteen Points will not be reviewed here. (They are listed in Section 1.6.) What I see as some of the most important and sometimes controversial lessons to be derived from the Points and Dr. Deming's papers and talks are: (1) Think in terms of processes, stability, common and special cause variation in going about your responsibilities, whether managerial or purely technical. (2) Managers should be leaders not supervisors. (3) Emphasize teamwork. (4) Do not look to slogans and goals as replacements for fundamental changes in systems that will give persons the chance to do what they generally wish to do, namely, "a good job". (5) Do away with short-term thinking, for example, concern for the statement of quarterly profits, and replace it with concern for long-term survival. (6) Create a constancy of purpose in regard to improvements in quality and productivity. (7) Do away with the traditional appraisal system that serves to differentiate among workers and is more counter productive than beneficial. Do away with grading in schools.

When you hear Dr. Deming, you hear a man deeply in love with America. You hear a man who is frustrated by what he sees as stupidity and ineptitude in regard to the management of her industry and in the rote and mindless application of statistics by her engineers, statisticians and managers. He is critical of the way statistics is presented in most textbooks and the way he sees it as being taught at the colleges and universities. He is not anti-intellectual; he understands that theory must underpin the methods we use; he says that, without it, persons are doomed to repeat past errors over and over again.

Dr. Deming is impatient with the slowness in America to adopt what he says. He knows the trends that are working against us; the fact that our leading exports are scrap iron and paper, that we are approaching colony standing relative to Japan in regard to sending raw materials and buying the manufactured products in return.

I believe that Dr. Deming has done more than any other person to raise the visibility of "statistics" and its proper application, and he has done it on a world-wide scale. He has been true and consistent in his insistence on proper professional conduct and ethics in its use. He has contributed to its theory and his practice has provided paradigms for all who wish to conduct and report scientific studies. We should be thankful that his long life has given us the opportunity to benefit from his tireless efforts to teach us; we should wish that the enthusiasm and hope that he has created will never ebb.

Section 1.6 Deming's Fourteen Points

Here we list the Points as given in Deming (1986, pp. 23, 24). He states:

"The 14 points apply anywhere, to small organizations as well as to large ones, to service industry as well as to manufacturing."

1. Create constancy of purpose toward improvement of product and service, with the aim to become competitive and to stay in business, and to provide jobs.

2. Adopt the new philosophy. We are in a new economic age. Western management must awaken to the challenge, must learn their responsibilities, and take on leadership for change.
3. Cease dependence on inspection to achieve quality. Eliminate the need for inspection on a mass basis by building quality into the product in the first place.
4. End the practice of awarding business on the basis of price tag. Instead, minimize total cost. Move toward a single supplier for any one item, on a long-term relationship of loyalty and trust.
5. Improve constantly and forever the system of production and service, to improve quality and productivity, and thus constantly decrease costs.
6. Institute training on the job.
7. Institute leadership (see Point 12). The aim of supervision should be to help people and machines and gadgets to do a better job. Supervision of management is in need of overhaul, as well as supervision of production workers.
8. Drive out fear, so that everyone may work effectively for the company.
9. Break down barriers between departments. People in research, design, sales, and production must work as a team, to foresee problems of production and in use that may be encountered with the product or service.
10. Eliminate slogans, exhortations, and targets for the work force asking for zero defects and new levels of productivity. Such exhortations only create adversarial relationships, as the bulk of the causes of low quality and low productivity belong to the system and thus lie beyond the power of the work force.
11a. Eliminate work standards (quotas) on the factory floor. Substitute leadership.
  b. Eliminate management by objective. Eliminate management by numbers, numerical goals. Substitute leadership.
12a. Remove barriers that rob the hourly worker of his right to pride of workmanship. The responsibility of supervisors must be changed from sheer numbers to quality.
  b. Remove barriers that rob people in management and in engineering of their right to pride of workmanship. This means, *inter alia*, abolishment of the annual or merit rating and of management by objective.
13. Institute a vigorous program of education and self-improvement.
14. Put everybody in the company to work to accomplish the transformation. The transformation is everybody's job.

Section 1.7 References

Boorstin, Daniel, *The Americans: The Democratic Experience*, Random House, New York, 1973.

Deming, W. Edwards, *Out of the Crisis,* Massachusetts Institute of Technology, Center for Advanced Engineering Study, Cambridge, MA, 1982, 1986.

Fowler, Elizabeth M., "University Heeds Advice on Management," *The New York Times,* Careers Section, March 7, 1989.

Halberstam, David, *The Reckoning,* Avon Books, The Hearst Corp., New York, 1986.

Katz, Donald R., "Coming Home," *Business Month,* 57–62, October, 1988.

Scherkenbach, William W., *The Deming Route to Quality and Productivity; Road Maps and Roadblocks,* CEEPress Books, George Washington University, Washington, D.C., 1988.

Walton, Mary, *The Deming Management Method,* Dodd, Mead & Company, New York, 1986.

Chapter

2

Enumerative Studies

In this chapter, we discuss the concept of an *enumerative study* and contrast it with an *analytic study*. Deming (1950, 1953) first used these terms in explaining an important distinction among statistical studies. We discuss the terms *population, frame, random sampling, statistical inference, sampling error,* and *measurement error* in connection with enumerative studies.

The above concepts are discussed in connection with my experiences in sampling utility poles and petition signatures. Along the way, I will discuss the concept of "random" in the connection with the way the Michigan Lottery improperly made its selection of finalists for a certain $1,000,000 drawing.

When I speak of "random selection of a single unit from a population", I will mean that each unit in the population is given *equal* probability of being selected. Equal probability is a relationship that can never be known to hold precisely since equality can not be proven. However, with some methods of selection it is a reasonable model with which to describe the probability distribution among various outcomes.

In Section 2.1, the concepts of population, frame, random selection, and simple random sampling are introduced. In Section 2.2, there is a further elaboration of these concepts together with a discussion of enumerative studies. In Section 2.3, examples of statistical inference in enumerative studies are given along with some discussion of different modes of selecting samples from a population, for example, cluster random sampling, systematic random sampling and stratified random sampling. In Section 2.4, the distinction between sampling error and measurement error is discussed. Section 2.5 contains a summary of sorts and some discussion of forecasting.

Dr. Deming has written extensively on what constitutes proper statistical design and practice. He has written on the presentation of statistical evidence and the proper role that a statistician must have in working with attorneys and subject-matter experts. (See Deming (1950, 1953, 1954, 1960, 1965, 1975, 1976, 1986).) I believe that his writings are the best source of clear exposition on what constitutes proper statistical practice. However, in this book I choose to use some terms that Dr. Deming believes are best left out of the statistical literature, for example, "population", and "statistical significance". See Deming (1975, p. 151).

I hope to give you a flavor of the excitement and suspense that accompanied some of my experiences in the application of statistics and the care that must be exercised in doing what appear to be simple enumerative studies. I was cross-examined at length by attorneys hired by the utilities in regard to the enumerative study of utility poles that I directed. This cross-examination was an attempt to weaken the impact of the findings. (Cross-examination is about as much fun as losing the seat off your bicycle on the first day of a twelve-day bicycle trip.) I am pleased that a close and constructive role in working with the Elections Division of the State of Michigan has led to the adoption of reasonable procedures for sampling petition signatures in Michigan and making inference about the validity rate. I am also pleased that the Michigan Lottery has learned to appreciate the nature of "random sampling" after using improper methods to select finalists for a drawing.

These experiences and many more have led me to appreciate the fact that concepts that many specialists in statistics may regard as simple and unworthy of much attention in books are, in fact, not so simple to the practitioners of statistics. Thus, I discuss and emphasize the simple but fundamental issues. Statistical studies are essentially worthless if accomplished without proper attention to the fundamental matters.

Section 2.1   Introduction to Population, Frame and Random Sampling

Consider a club consisting of six individuals, whom we will label as A, B, C, D, E and F. One member is to be chosen at random to go to the store. Here we will regard the Club as a *population* of individuals and the selection of a member of the Club as a random selection of an element from the population. What is a reasonable way to make this selection? There are many reasonable ways.

Example 2.1   Dice

The Club indulges in dice games and has a well-made die in its possession. This die is a cube made of plastic with a uniform mass distribution, that is, it is balanced. We call it a fair die. The members agree, before the roll of the die, that A goes to the store if a 1 is rolled, B goes if a 2 is rolled, . . . Below we give the complete correspondence of this population with what we will call a frame.

Frame:	1	2	3	4	5	6
	|	|	|	|	|	|
Population:	A	B	C	D	E	F

Here a mechanism exits to select a number at random from the numbers 1, 2, 3, 4, 5, and 6. These numbers constitute a *frame*, and we see that each element of the population has been identified with one and only one element of the frame before the sample is drawn. With the roll of the die, an element from the *frame* is drawn and this determines the selection of an element from the population. ■

Example 2.2   Coins

Suppose the Club does not have the balanced die, but it does have a coin available to toss. In this example, we will learn how to use the coin to select a member at random from the Club. The coin is balanced, giving equal probability for the outcomes "Heads" and "Tails", hereafter denoted by H and T. We call this a fair coin. By placing the coin in a can, shaking the can, and then tossing the coin from the can, a H or T will be observed. By repeating this operation two more times and keeping track of the outcomes, a sequence of H's and T's will be observed. We will assume that the probability for H on each toss is the same (1/2) regardless of the outcomes of preceding or future tosses. These are called independent tosses of a coin. The eight possible outcomes for n = 3 tosses are listed below in a set called a sample space; each outcome occurs with probability 1/8.

$$S = \{TTT, TTH, THT, HTT, THH, HTH, HHT, HHH\}$$

A tree diagram that describes this experiment with the coin is shown in Figure 2.1. Tree diagrams are very useful in describing simple experiments that evolve in a sequential manner.

By performing this experiment, we pick an element at random from the set S. We can use S as a frame for our population of club members. We identify A with the element TTT of the frame, B with TTH, . . . resulting in the correspondence shown below.

Frame:	TTT	TTH	THT	HTT	THH	HTH	HHT	HHH
	|	|	|	|	|	|		
Population:	A	B	C	D	E	F		

There is an excess of two frame members after we complete the identification, but not to worry. It can be shown with a more advanced analysis of probability, that if the experiment is repeated independently until one of the first six listed outcomes occurs, then each of these outcomes has probability 1/6; thus, an element is determined at random from the population. ■

In general, the frame is the list or set to which is applied the random selection mechanism. The identification of the population with a subset of

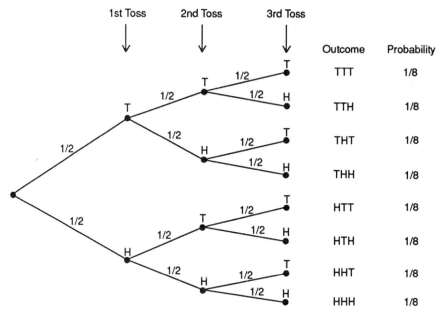

Figure 2.1 Tree Diagram for Three Independent Tosses of a Fair Coin

the frame translates selections from the frame to selections from the population.

With Examples 2.1 and 2.2, we have illustrated how to select a single club member at random to go to the store. If the trip will require two club members, how can the two be selected at random? To select a set of two elements at random from a population, simply identify one element by random selection, and then independently repeat the process of random selection from the population until an element different from the first is selected. (More advanced analysis shows that the method thus described gives equal probability of selection for every subset of size two of the population.) In the case of using the frame consisting of the outcomes from the roll of the fair die, the die is rolled to determine one of the two to go, and then rolled again and again until a second club member is determined. For example, the outcome from this experiment 1, 4 determines that A and D will go, and the outcome 3, 3, 2 determines that C and B will go.

Selecting three at random to go to the store requires taking the idea one step further. Implementing this with the die, the result 4, 2, 4, 2, 6 determines that D, B and F go to the store. Of course, the method extends to allow for the selection of any number n (up to six) of the club members to go to the store.

We call the method of selecting a subset of n elements in the manner described above *simple random sampling* of n elements. With it, all subsets of size n in a population of size N are given equal probability $1/C^N_n$ of being the subset selected. (Here and throughout this book, the symbol C^N_n

denotes the number (N!)/[n! (N-n)!], which is the number of combinations of N elements chosen n at a time.) With simple random sampling, all individual elements in a population of size N are given the same chance n/N of being selected into the sample. The use of simple random sampling for selecting a member of the Club to go to the store was to assure that the individual members had an equal chance of being selected. In selecting two to go, it assures that every one of the $C^6_2 = 15$ subsets of two individuals from the Club has equal probability 1/15 of being selected.

The use of simple random sampling with the Club was to achieve fairness in the selection process. If in selecting two to go to the store the outcome is the two male members of this six person club, the selection process remains fair, but the outcome may be considered to be not fair. This distinction is extremely important; it suggests that agreement be made in advance of the selection as to the process to be used in the selection and that agreement be made in advance that there be adherence to the outcome. This comment applies in regard to random selection for fairness, and, arguably, for random selection for the purpose of making inference.

As an example of an unusual use of simple random sampling, consider the seldom-used Michigan law that calls for the random removal of excess paper ballots from election ballot boxes where the number of ballots exceeds the number of persons listed in the poll books as having voted. (Paper ballots are seldom used any more having been replaced by voting machines and punch card devices.) Without evidence as to which candidates(s) may have benefitted from the excess, random removal seems fair. Proportional reduction of the counts is another fair method, but it may lead to non-integer results for counts. Discussion of this and much more in regard to remedies for election irregularities are found in Gilliland and Meier (1986).

Protocols in the use of statistics are extremely important and today there is more and more interest in such as statistical theory and methods play increasingly important and visible roles in law and decision-making. In this regard, I like the work by Finkelstein (1978, Chapter 7) as well as the writings of Dr. Deming. Protocols can be used to help assure that data have proper and agreed upon pedigrees, and that challenges be reasonable and structured along certain lines.

In Section 2.3, we will learn of the importance of random sampling in regard to statistical inference. With random sampling, it is possible to quantify the precision of estimators based on the sample so that here the use of random sampling is not related to a fairness issue per se.

Dice and coins and slips of paper in a hat will not take us very far in selecting elements at random from the populations that we will face in the practice of statistics. However, many software packages exist which seem to do a reasonable job of simulating the selection of single elements and subsets of elements from sets of integers and continuous distributions. In Chapter 4, I will have occasion to use a software package to simulate independent observations from distributions to illustrate various points. Before using such a software package in a serious application, be sure to check the literature for reviews and reports of the adequacy of the random number generator being implemented and the manipulations being used to simulate the sampling from specific distributions such as uniform, normal. . . .

Quite often in practice, a frame is used that is larger than the population, as in Example 2.2. We now give an example of an actual experience involving this concept.

Example 2.3 Random Sampling of Petition Signatures

This example concerns the random selection of signatures from the set of 319,468 signatures counted to have been submitted in a state-wide petition drive for a constitutional amendment in Michigan in 1978. These signatures appeared on 27,277 petition sheets; the sheets were given the numbers from 1 to 27,277. Each sheet had 15 lines that were numbered 1 through 15. The average number of signatures per sheet was 11.7.

How can a simple random sample of 500 signatures be selected from the 319,468 appearing on these sheets? We know that this can be accomplished by selecting one signature at random after another from the population of 319,468 until 500 different signatures have been selected. However, there is no simple list of the signatures. (Think about how you would select one signature at random to realize that this is not a simple task.)

Fortunately, the signatures do appear in an organized way, that is, on a subset of the 27,277 × 15 = 409,155 lines on the 27,277 numbered sheets. We can use the list of all 409,155 lines as the frame.

A single signature can be selected at random from the population of signatures by selecting an integer I at random from 1 to 27,277 followed by an integer J at random from 1 to 15 to determine a sheet I and a line J on that sheet. If line J of sheet I has a signature on it, that signature is the randomly selected signature. If not, the process of selecting (I,J) is repeated until a signature is identified.

The above process can be repeated over and over until 500 unique population items are selected. In summary, in this application we can use the list of all 27,277 × 15 = 409,155 lines as the frame. The subset of lines that contain signatures is identified with the population of signatures.

This all seems pretty simple. Maybe this is not so simple. In the actual experience in 1978, I discovered that a consultant hired by the State of Michigan had not selected a simple random sample of 500 signatures as he had been asked to do. The individual was a computer scientist who did not have a clear understanding of something apparently as simple as the selection of a signature. In fact, he selected a signature "at random" by selecting a sheet at random from the 27,277 sheets and then selecting a signature at random from the sheet thus selected. Figure 2.2 is a partial tree diagram that shows that this process gives 15 times greater probability of selection to the

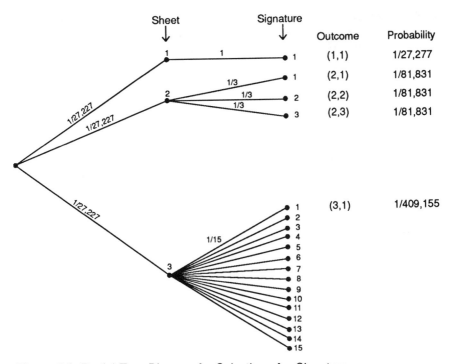

Figure 2.2 Partial Tree Diagram for Selection of a Signature

signature appearing alone on Sheet #1 as compared to any one signature appearing on Sheet #3, which contains fifteen signatures.

We will pick-up on this example in the next section where I will tell you more about the ramifications of this error. The reader can refer to Gilliland (1981) for more details on experiences with sampling and inference for various initiative, referenda and constitutional amendment petitions in Michigan. ■

Exercise 2.1. (a) Use a tree diagram to determine the probability that the experiment of rolling a die has to be performed only twice to select two members from the Club to go to the store. (b) Use a tree diagram to determine the probability that the experiment of tossing a coin three times has to be performed only twice to select two members from the Club to go to the store. (c) Use a coin and the coin tossing experiment to actually pick two members of the Club to go to the store. Keep track of the number of times you had to run the experiment in order to make the selection of the two club members. Compare results with your classmates. ■

Exercise 2.2. On December 30, 1981, the Michigan Lottery selected five finalists from among 11,561 winners of $50 in the instant lottery game "Three Aces" for a $1,000,000 drawing to be held later. I will describe the method used by the Lottery that evening to pick a single finalist "at random". You will be instructed to determine the probabilities of certain outcomes.

The lottery used as a frame the set of 11,561 five-digit numbers

$$S = \{00001, 00002, 00003, \dots, 11561\}$$

after having assigned one of these numbers to each of the 11,561 contestants. (Here I am giving a simplified version of what in fact was the case to keep the analysis simple. In fact, there were fewer than 11,561 winners in the pool for the selection of finalists, but S was used as a frame. Details can be found in the Lansing State Journal article by Keith Gave (January 17, 1982) where the analysis of Professor Martin Fox is described.) By using now-familiar devices that blow numbered pingpong balls around and allow that one be drawn off at the push of a button, an element was selected from the frame S. The Lottery officials used the following procedure. First, a device with two balls labeled 0 and 1 was used to select the leading digit in the five-digit number. If the outcome were 0, then a device with ten balls labeled 0, 1, . . . ,9 was used to select the second digit; if the first outcome were 1, then a device with two balls labeled 0 and 1 was used to select the

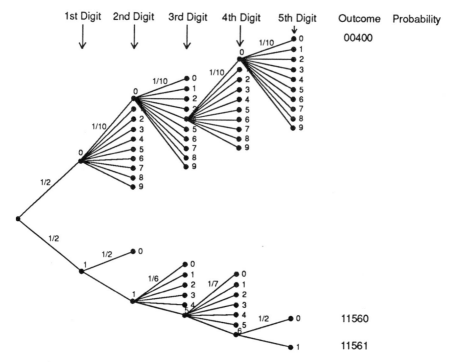

Figure 2.3 Partial Tree Diagram for Selection of a Finalist

second digit. (Note that there are no elements of S that start with the digits 12, 13, 14, 15, 16, 17, 18 or 19.) The partial tree diagram in Figure 2.3 gives some of the probability structure for the method used by the Lottery to select a single finalist.

Mr. X was assigned the frame element 00400. He was the person who raised the issue of unequal probabilities of selection. Mr. Y was assigned the frame element 11560. (a) Use the multiplication law for probabilities to show that Mr. X is selected with probability 1/20,000 and Mr. Y is selected with probability 1/336. (b) Determine the probability that 00001 is selected. (c) Determine the probability that 11098 is selected. (d) Design a selection system using devices with numbered pingpong balls that gives 1/11,561 chance of selection for each element of S. You should strive for simplicity in your system. You should design one that can provide a visual display of what is going on for an audience of spectators. ■

Postscript to Exercise 2.2. You might be interested to know that the Commissioner of the Lottery in January 1982 apparently appealed to law and directives that do not strictly define "random" and to other arguments before reacting in a strange way to this unfortunate situation. He was quoted in the article referenced earlier as saying that "It is my belief that the drawing was fairly conducted and that each of the contestants had a fair chance of winning." I believe that this is a pretty stupid reaction given the above facts and the importance of maintaining a good public opinion of the Lottery. A second drawing for finalists should have been held using correct methods and a second $1,000,000 winner determined. ■

Persons seem not to be able to use their minds to select elements *at random*. In one large class of N > 200 students, I asked that each use his/her mind to select one of four columns at random from a page of an open campus telephone directory. I will label the four columns from left to right on the open page as 1, 2, 3 and 4. The students chose 2 (and 3) a much larger proportion of times than the expected N/4 times. This confirmed earlier findings from the same experiment involving a different group of persons. I believe that, generally, persons have selection bias in favor of 2 and 3 in using their minds in an attempt to simulate random selection from {1, 2, 3, 4}.

Persons seem not to be able to use their minds to simulate *independent* random selections. In a talk at Michigan State University many years ago, Professor P. Erdos told of how he asks students in his classes to simulate a large number of independent tosses of a fair coin. He told of how the students tended to produce outcomes that were much more regular than is likely under truly random generation. Specifically, the students tended to put down a pattern, where at each stage, there was a fairly even number of H's and T's, thus, avoiding long runs of H's and long runs of T's.

The fact that independent random events produce less order and regularity than people expect is reported by Thomas Sowell in his article

"Ideas from Statistics Don't Always Add Up," *Detroit News*, July 23, 1990. He states:

> "A statistics teacher used to play a little game with his class. He would give them some simple experiments to do while he was out of the room. They were supposed to write down the numbers that came out of these experiments. They were also told to write down another set of numbers, made up out of their heads.
>
> When the instructor returned to the classroom, he was supposed to guess which set of numbers was real and which set was phony. Although the students tried their best to fool the teacher, he would usually just glance at the two sets of numbers and immediately tell them which set was real.
>
> The students were baffled as to how he could tell. After a while, he finally let them in on the secret: The numbers they made up were too even, too orderly, too stable. Real statistics from the real world are seldom like that.
>
> Real statistics are not even or orderly. They are irregular. They jump around or are lopsided."

Persons can not be expected to simply use their minds to make independent random selections. We have too many misconceptions and biases in our understanding of what "at random" is and what "independent" is. Therefore, we are well-advised to use physical devices, random number tables and software packages rather than our minds alone to perform independent random selections.

There is large and important literature dealing with human judgment under uncertainty and the perceptions and understandings of the concepts "at random", "probability" and "risk". Look under the name Amos Tversky in gaining entry to this literature.

Section 2.2   More on Enumerative Studies

An *enumerative study* concerns either enumeration or inference in regard to a fixed population of items. An enumeration is a counting process applied to perform a complete census of every item in the population, and an inference is a statement about the population based on sample information. The focus of the study may be to determine or estimate a frequency or proportion of items in the population that have a certain characteristic; hence, the adjective enumerative. An example would be the determination of the number of records in a large data base which have at least one field with a data entry error. The focus of an enumerative study may be to determine or estimate a mean characteristic for the population, for example, a mean income or mean dollar value.

Deming (1950, 1953) contrasts studies of this type with those that focus on the reasons and underlying mechanisms that produce the proportions or means; he calls the latter type *analytic studies*. The distinction is based in part on the purpose and the focus of the statistical study. The distinction must be recognized when designing a study and when assessing the precision of statistical estimates.

In an enumerative study, the entities making-up a population are fixed, well-defined entities. Any entity could be determined to belong or not belong to the population based on the application of the defining conditions for membership in the population. At this level of conceptualization, the population is no more than what is called a "set" in mathematics, and, at this level of abstraction, a population is often depicted by a Venn diagram of a set.

A frame is a list that has a well-defined subset that is in one-to-one correspondence to the elements or sampling units of the population. A simple list of the integers 1 through 100 will serve as the frame for a population consisting of 100 entities if each of these integers is identified with one and only one element of the population. As we have seen, lists that are more extensive than the population of interest can also serve the purpose of a frame.

Sometimes a frame is next to impossible to construct for a population. There is not list or frame for all utility poles in Michigan to which are attached cable television lines. We will discuss this example later when we tell how we functioned without a frame and used a two-stage sampling plan involving cluster sampling as the method to select a sample of 600 utility poles for investigation.

Dr. Deming insists on *operationally defined* concepts and lots of care is required in creating these definitions. See Deming (1953, 1986). Basically, a concept or term is operationally defined if a set of defining relationships or tests are given for the concept. These tests must be ones that can be applied in an unambiguous manner. In practice, there is a tendency to use terms loosely in the haste to get to the meat of an application. Such haste often is accompanied by lack of attention to definitions; the study is then like a house of cards, one very prone to topple into disarray.

One might speak of "the population of all households in East Lansing". However, these words do not define a population. Exactly what are these households? Are they persons, buildings, residences, apartments, groups of persons, mutually exclusive groups of persons, . . . ? After attempting to answer the question for yourself, you might decide at what point in time the definitions are to apply. On July 1, 1990 at noon? Your definition applied at two different points in time will surely lead to different populations since East Lansing is part of a dynamic and changing world. Therefore, you must specify exactly what time your definition is to apply or simply realize that the result of the application of your definition will change with time.

The concept of "household" may be too difficult for us to deal with here. Let me take a simpler example with which I have intimate familiarity. I was asked to conduct an enumerative study in regard to utility poles in the State of Michigan to which were attached cable television lines. An issue had arisen as to a fair rent that utilities, the owners of the poles, could charge the cable television companies who use the poles to carry their cables to the vicinities of the homes and apartments of their subscribers. Although the issue that was joined before the Michigan Public Service Commission concerned a particular utility and a particular cable television company, it was decided that the enumerative study should involve the population of all utility poles in Michigan that carried cable television lines. Realizing that some poles were being added to the set each day and some poles were being replaced each day, it was clear that we had to specify a time in our definition for our population. We chose to define our population as all utility poles in Michigan that were owned by utilities and were being rented to a cable television for the purpose of carrying a television cable as of a specific date. Since records were kept by the utilities of their physical plant and their rentals, for any utility pole in Michigan at any time, it could be determined whether the pole was a member of the population or not. Admittedly, there still remain discrepancies and ambiguities that might have to be resolved in applying this definition, but such would be small issues relative to the issues that remained to be settled in this study. Here we are insisting upon the careful use of language in defining concepts such as population and in giving persons tests for determining whether an entity is in the population or not. There is little sense in proceeding with enumeration or with sampling and inference unless first there is this concern for basics.

Back to the utility poles. The physical characteristics of the poles that were of interest and importance in setting rental rates were determined and defined. One characteristic called "usable" space was based on various measurements including the height of the top of the pole above the ground level. What is ground level? Have you seen the top of a utility pole? Do you realize that they are often cut at an angle, I presume so that they will shed rain water. Operational definitions were created for the concept of the height of the top of the pole above ground level, and the survey crew that was to collect the data was trained in applying those definitions.

Let's consider an even simpler enumerative study. Effective October 1, 1973, a law came into effect in the State of Michigan providing for no-fault insurance coverage for registered motor vehicles. In response to legal challenges, the question arose as to the percentage of automobiles registered (licensed) in Michigan that were without liability coverage. Professor Leo Katz was asked to direct the study. (See Katz (1975).) The population was defined as the collection of all passenger vehicles that were registered for the road as of the day of the sampling. Each such vehicle had a unique registration number and, supposedly, was listed on the computer tape of all registered vehicles maintained by the Michigan Department of State.

In this case, the computer tape provided a frame for the population. Professor Katz employed a two-stage systematic random sampling plan to select around 250 registration numbers from a population of over 4,000,000 in the computer tape file. In the next section, we will discuss the use made of the sample information. You might think that the sample he chose was too small to provide sufficiently precise inference in regard to the population. Since Professor Katz anticipated that it would take private investigators to determine the insurance status for some sample vehicles, he kept the sample small and manageable yet with size sufficient for the precision required in the application. This strategy helped to minimize the chance for measurement error in the sample. Measurement error must always be anticipated and controlled.

Section 2.3   Statistical Inference in Enumerative Studies

In the case of the population of 319,468 petition signatures discussed in Example 2.3, the sample of 500 was selected so that information would be provided on the validity rate for the signatures in this population. The Michigan State Board of Canvassers was to act on the question of the certification of the petitions as to sufficiency of number of valid signatures on the evening of August 28, 1978 after being presented with the results of investigations of the 500 sample signatures. The petitions concerned a controversial tax issue. Emotions were running at fever pitch as the State was in the throws of a tax-payer rebellion of sorts. The night was very warm and with the windows open one could see a commotion. A person in a gorilla suit was urging the public to honk horns in favor of approving the petitions as to sufficiency of valid signatures. Perhaps, cold, hard statistics were to determine the fate of the petitions and not the mob. The Elections Division staff was oblivious to the fact that the computer scientist had not selected signatures at random, potentially biasing the sample, and all was to be decided at this meeting involving concerned and volatile citizens, Board members appointed by the Governor, the Director and technical staff from the Elections Division, and, of course, dozens of attorneys.

State of Michigan law specified that the petitions in Example 2.3 must contain at least 265,702 valid signatures in order to qualify the proposed constitutional amendment for the ballot in the general election of Fall 1978. (This figure is adjusted following each gubernatorial election and is, by law, a specified percentage of the total number of ballots cast in that election.) Therefore, the minimum validity rate required for the petitions under consideration was 265,702/319,468 = 83.17%. The Elections Division staff had initially determined that the *sample* validity rate was 399/500 = 79.80%. Based on the theory of simple random sampling, the standard error or estimated standard deviation of this estimate was 1.79%. However, the formula upon which this standard error was calculated is justified by the simple random sampling model. It is *not* appropriate for the method of selection that was actually used that gave probability proportional to size (of petition sheet) for selection of individual signatures into the sample. As best I could determine and as reported in Gilliland (1981), the method of selection used by the computer scientist and the sample determinations provided for an unbiased estimate 80.84% for the population validity rate with a standard error 1.92%.

On August 28, 1978 interested parties were able to demonstrate to the Board Canvassers that the Elections Division staff had made a fairly large number of errors and questionable decisions in the determination of the validity status of sample signatures. Following this exposé, the Board voted to throw out the sample, and it directed the Elections Division to sample again and to use higher standards of quality control in processing the new sample signatures. The Board did not want to see a high level of measurement error in the processing and in the investigations of the new sample signatures. Based on what the Board knew, it should have acted to retain the sample and to have the measurement error removed from it. Based on what I knew about the way the sample had been drawn, I thought it best to throw out the sample and start fresh. Therefore, I said nothing at the meeting and the error by the consultant was not exposed that night. In private, I told the consultant and the Elections Division of the error, and it was corrected before the selection of the new sample of 500.

The key lesson here is to always examine the way in which a sample has been selected and to not blindly trust individuals to have necessarily carried out the sampling according to the specified plan. Another thing to note is that formulas in books based on simple random sampling are inappropriate and possibly seriously misleading if the sample was not selected as a simple random sample giving equal probability of selection to every subset of size n. The pedigree of the sample is critical to understanding its precision and possible bias.

In the utility pole survey, I did not use simple random sampling; rather, I used a two-stage sampling plan involving cluster random sampling. The first stage involved the selection of a local franchise of a cable television company, and the second stage involved the independent random selections of six clusters of four somewhat neighboring poles from the poles rented by the franchise. This entire selection process was repeated independently 25 times to produce a sample of 150 clusters of 4 poles each. Based on the prescribed method of sampling, I was able to testify that an estimate for the average usable space across the population of poles was 16.4 feet with a standard error of 0.5 feet. Because I had used an estimated size for each cable television franchise in the selection process, and because I could select clusters with only approximate equality of selection, it was essential that I report the results of an error analysis that quantified the possible bias introduced by my methodology. I determined and reported that the possible bias in my sample estimate was very small compared to the standard error.

The standard error is important as a measure of the precision of an estimate. With certain models and distributional assumptions, an unbiased estimator ± (2 standard errors) is an approximate 95% confidence interval estimator. In the utility pole experience, 16.4 feet ± 1.1 feet could be taken as a nominal 95% confidence interval estimate for the value of the critical population parameter.

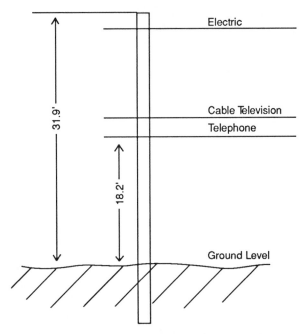

Figure 2.4 Sketch of a Utility Pole Showing 13.7′ of Useable Space.

The estimates and the data collected in the sample survey suggested that the number of utility poles in the population of interest was over 1,000,000. I will now discuss some of the considerations that led to the sampling plan and the sample size of 600 poles.

The key population parameter or characteristic of interest was the average length μ of utility pole from the lowest line attached to the pole to the top of the pole. The lowest line attachment is either the cable television line or a telephone line if a telephone line was attached to a pole. Figure 2.4 is a sketch of a pole which has three attachments; the depicted pole has 13.7 feet of length above the lowest attachment.

The cable television industry believed that in Michigan the population average μ was at least 13 feet. The distance from the lowest line attachment to the top of a pole is referred to as usable space in a certain economic model for apportioning costs for a pole to users of the pole. The figure 13 feet for usable space was important for certain reasons, one being that a federal guideline for rental charges for utility poles was predicated on there being on average at least 13 feet of usable space on a pole.

Based on discussions with experts on the physical plants of utilities and the results of a survey of utility poles in Ohio, it was thought that the Michigan population average μ was between 14 and 17 feet. (This interval was based on judgment and opinion, and the confidence that this interval contained μ could not be quantified. However, as we shall see, this preliminary thinking helped us in selecting a sampling plan and a sample size.) Thus, we believed that the population average differed from the 13 foot benchmark by at least 1 foot. For the purpose of presenting reasonably precise estimates for the Public Service Commission to demonstrate the difference, we decided that the standard deviation of the sample estimator should be 0.6 feet or less.

I thought that the usable space in the population would range from about 0 to 40 feet. (Surprisingly, there are a relatively few short poles in Michigan carrying only a cable television line and owned by utilities. Such may have 0 usable space. There are a relatively few 65 foot poles in Michigan; they have bases about 9 feet below ground level and lowest attachments about 18 feet above ground level, leaving about 38 feet of usable space. Most of the population poles were thought to be from the 35–45 foot classes.) The range for the population distribution of usable space and my understanding of how a population standard deviation relates to shape characteristics of different distributions in a variety of differently shaped distributions suggested to me that the population standard deviation σ of usable space was no more than 9 feet. (It is never more than one-half of the range of the population.) See Figure 2.5 for some examples involving various distributions. Since with simple random sampling, the standard deviation of the sample average is $\sigma/\sqrt{n}$, we see that n = 225 poles chosen as a simple random sample from the population would suffice if the pre-sample guess on a bound for σ were accurate.

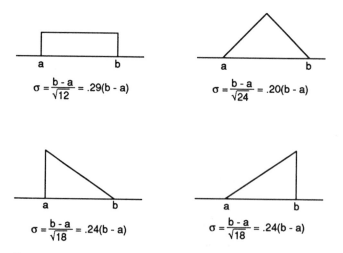

Figure 2.5 Relationship of Standard Deviation σ to Range for Certain Continuous Distributions

There were several difficulties with using a simple random sample of n = 225 poles. I could not figure out how to select 225 poles at random from the over 1,000,000 in the population. There was no frame or list of these poles available to me. The utilities who owned these poles were not cooperating with the cable television industry that was sponsoring this study; the two were adversaries. My experience with pre-sample meetings with the utilities showed that there was no chance for cooperation. (This is a case where the decision-making body, here the Public Service Commission, should be playing the lead to be sure that relevant statistical information is developed in an independent and mutually agreed upon manner.) Besides, the selection of a simple random sample of 225 poles would likely result in the survey crew having to visit many fairly widely dispersed locations in the State of Michigan to measure the sample poles. This would result in prohibitively large travel costs and time.

I decided to use the two-stage sampling plan which would likely concentrate the sample poles geographically and still result in the necessary precision. I decided to use the following procedure to select a pole. First, a cable television franchise from a list of 102 operational in Michigan is selected. Then six poles are selected at random from those rented by the chosen franchise, which we will call "starting" poles. Here the franchises make-up the primary sampling units (PSU's), and the poles make-up the secondary sampling units (SSU's). In fact, I decided that this procedure would be independently repeated 25 times and that 3 poles near each of the ones chosen would be surveyed in addition to the 150. I specified in an unambiguous way how to determine the 3 poles to be surveyed along with each starting pole. Using directions from a compass provided the survey crew and the cable line (lines) attached to the starting pole, an operational definition was given for completely determining the cluster of 4 poles, including the starting pole. In the final analysis, the sampling plan that was used was a two-stage plan with the PSU's being franchises and the SSU's being clusters of 4 poles.

We see that a sample size of 600 poles is specified by the above sampling plan. Some theory, the fact that n = 225 was determined to give sufficient precision under simple random sampling, an understanding of intra-cluster correlation and its likely magnitude in this application, economic considerations related to the cost of collecting the data from the sample poles, and the fact that no frame existed for the entire population led me to choose the plan with 25 PSU's followed by 6 clusters of size 4 each. Other two-stage sampling plans were probably as good; we selected this one.

I will not go into more details of the plan except to say that the PSU's were not chosen at random from the list of 102 franchises, but rather at random with probability proportional to an estimated size of the PSU, where size is the number of population poles in a franchise. If the estimated sizes were correct, this method of sampling would give equal probability of selection to each cluster of 4 poles in the population. This makes the cluster sampling what is called "self-weighting" and makes the analysis of the sample data very simple. I could regard the 25 PSU averages of usable space as a random sample of unbiased estimators taken with replacement. The average of these averages is an unbiased estimator of μ, and the standard deviation of these averages divided by $\sqrt{25}$ is its standard error. By the way, as stated earlier, the estimate for μ turned out to be 16.4 feet with a standard error of 0.5 feet. The nominal 95% confidence interval estimate of μ was 16.4 feet ± 1.1 feet, providing very strong evidence that μ was at least 13 feet.

The sampling plan had lead to an estimate with the precision that was sought. It had concentrated the 150 clusters in only 19 different local franchises with one large franchise (Grand Rapids) having 18 clusters for a total of 72 poles. The travel and time costs were acceptable and the crew performed wonderfully. I had no hesitation in regard to testifying in regard to this sample evidence because of the quality checks and training involving the survey crew.

As I alluded to in Section 2.1, it was difficult to construct frames for the poles. For some sample franchises, I had to use sampling of subscriber addresses and poles located near those addresses to select poles; in other PSU's, I had to use random sampling of locations (coordinates) and prescriptions for locating poles. (Point sampling is used to sample trees in a forest.)

Because of the two-stage nature of the sampling plan and the use of local franchises as PSU's, I needed to deal with frames for only a limited number (19) of areas in Michigan. Had I used a plan where, say, 150 clusters were chosen at random from the population of all poles in Michigan, it would have been necessary to deal with a much more complex frame problem and the data collection costs would have been prohibitive.

I kept a complete journal of all the decisions I made and all of the details of the sampling plan, the frames, and the random numbers generated to select PSU's and starting poles. My error analysis to determine the extent of the possible bias that was introduced by using estimated sizes in the selection of PSU's rather than actual sizes was expanded to include the effect of unequal probability of selection of starting poles at the second stage of the two-stage selection process. All of the details were given in direct testimony and under cross-examination before the Michigan Public Service Commission on November 12, 1982.

I learned several lessons from this experience in regard to the application of statistics. You should keep them in mind if you ever conduct a statistical study and the evidence is to be presented in an adversarial setting. First of all, read Deming (1954) as I did before I planned the utility pole survey.

Keep a good journal of your actions, decisions and observations as any scientist would do in conducting a careful experiment. Such may become fodder for the adversaries and will provide them with countless questions to use in the cross-examination, but the benefit from the honest and ethical approach to the application of statistics far outweighs the downside.

You will be well-served in directing a statistical study to get as involved as time permits with the data gathering portion of the study. Then you can testify as to whether the data were gathered according to your plan. With the utility pole survey, I traveled with the crew to several sites and observed the work. Based on what I saw and the training and detailed directions that I had provided, I was very satisfied with the work of the survey crew.

I even provided the crew with dice and directions of how to choose between poles if there was any ambiguity in the selection and location process for a starting pole or another pole in a cluster of poles. After the survey was complete, the crew reported that all 150 starting poles were easily located, but that one substitution for a prescribed sample pole had to be made with the roll of the die to determine which of two poles was substituted for it. It seems that the prescribed pole had an angry dog tied to it, thus preventing the crew from making the necessary measurements.

I kept the survey crew and the local franchise personnel in ignorance of what outcomes were favorable to the cable television industry so as to lessen the chance of bias in the measurement and location processes. One person from a local franchise who was going to help the survey crew locate the prescribed starting poles in his franchise, suggested to me that he knew where the tall poles were after getting the idea that tall poles were "good". I assured him that random methods were being implemented to select the sample poles, that we would not require his judgment in the selection process, but that we would require his help in locating the chosen sample starting poles in his franchise.

Are there people who are unethical in the practice of statistics? I believe so based on what I was to discover with this experience. While preparing for the Michigan survey, I had placed a call to the so-called statistician who had directed the Ohio survey of utility poles. I had read his report, felt that he had possibly overstated the precision of his estimate, and asked to see the raw data. He rather nonchalantly mentioned that he quickly destroyed the raw data after gathering it, and he proceeded to tell me that he could provide me with only the summary statistics. I am convinced based on reading his report and the telephone conversation that he gathered the data in cluster form but analyzed and assessed its precision using formulas appropriate for simple random sampling of individual poles. His testimony in Ohio had been erroneous, and, either on the basis of ignorance or unethical practice, he should be disqualified from such work. (As of this writing, there is no certification for statisticians and little chance that standards can be set, let alone monitored.)

Deming (1954, 1965) states a very strong position on ethical and professional standards in regard to the application of statistics. Meier (1986) is a lucid and interesting article on the subject.

The Katz (1975) study of the insurance status of automobiles registered (licensed) in the State of Michigan as of the day of the sampling in 1974 is a classic example of the proper use of statistics in an enumerative study. The study led to a 95% confidence interval estimate for the population percentage p of automobiles that did not have proper liability insurance. Specifically, the "exact" 95% confidence interval was calculated from the sample to be 2.0% < p < 6.5%, and the judge ended quibbling by stipulating that in fact p was in this interval.

Professor Katz selected a systematic random sample of about 250 cars from the computer tape file (frame) that listed all registered automobiles and trucks in Michigan. The vehicles were listed by license tag number which put the numbers in some sort of geographic order. Professor Katz selected a random position within the first 1,126 of the ordered listing and sampled that position and every 1,126th thereafter throughout the file. He followed this by subsampling the first sample using a sampling interval of 16. The sampling intervals had been chosen in advance to produce a sample of about 250 automobiles. Because of the fact that insurance status was known to be correlated with geographic location within the state, the systematic random sampling produced a sample that is very likely to be more accurate for estimating p than one selected as a simple random sample. Thus, Professor Katz could testify that, in his opinion, the confidence interval calculated from formula appropriate for simple random sampling had more than the nominal 95% level of confidence.

Professor Katz describes in some detail how the statistical evidence was presented and what arguments were mustered against it. Most of this sort of quibbling would be prevented if the decision-making body and the adversaries would agree in advance to the details of a study and would agree to abide by the results provided the study is performed according to acceptable standards.

I have not talked about hypothesis testing and statistical significance in this section since I believe that confidence interval estimation should be the method of inference used in most enumerative studies. The decision problem involving the certification of petitions can be phrased in the context of an hypothesis test, but this is not necessary for our development. Gilliland (1981) uses the language of decision theory to describe the operating characteristics of the decision rules described therein. However, much of the discussion of probability of error can be replaced by discussion of precision of estimators.

The language of hypothesis testing is rather arcane and convoluted. However, I do believe that it is instructive to consider the concept of statistical significance, and I introduce it rather informally in Chapters 3 and 4.

Section 2.4 Measurement Error

Sampling error refers to the difference between a value calculated from the sample and the value of the population parameter for which it is an estimate. The first value may be referred to as a sample statistic. For example, consider a population of six male children {A, B, C, D, E, F} with respective heights in inches

$$S = \{36.2, 39.0, 35.3, 34.9, 32.6, 35.6\}.$$

Here we can determine that the value of the population mean μ is 35.6". Suppose a sample of size n = 2 is drawn in order to estimate the mean height μ of the population of children. Suppose the sample consists of the children A and B. The sample mean is $\bar{x}$ = 37.6" so that the sampling error associated with this particular sample is 2.0".

Measurement error refers to an additional component of error that is due to the measurement process itself. For example, suppose that the sample heights determined by the person carrying out the study were taken with a measuring tape that had .5 inches removed from its lead end. Then this person would determine the heights of the sample children A and B to be 36.7" and 39.5" for an average of 38.1". Thus, the difference between this value and the population mean μ is

$$
\begin{aligned}
\bar{x} - \mu \quad &= \quad 38.1\text{–}35.6 \\
&= \quad (38.1\text{–}37.6) \quad + \quad (37.6\text{–}35.6) \\
&= \quad 0.5 \quad\quad\quad + \quad 2.0. \\
&= \quad \text{measurement} \quad + \quad \text{sampling} \\
&\quad\quad\quad \text{error} \quad\quad\quad\quad\quad \text{error.}
\end{aligned}
$$

Here the measurement error is a bias which adds .5" to each value determined by the measurement process.

Sometimes the measurement process introduces a component of variation and not simply an additive error component or bias. For example, suppose a proper measuring tape is used, Person U performs the measurement on Subject A, and U is prone to measure on the high side; perhaps, U always rounds measurements upward when unsure. Suppose that Person V performs the measurement on Subject B and V is prone to round measurements downward when unsure. This measurement process for the sample might lead to the reported sample heights 36.3" and 38.9". In this case, the two measurement errors +.1" and –.1" compensate for each other in the averaging process. The effect of these measurement errors does showup in the sample standard deviation.

Note that in both cases, the measurement error process will have an effect on the results if the study were a complete census of the population. With a complete census in an enumerative study, there is no sampling error; however, if measurement error exists, it will showup in the measurements taken in the census.

We see that there must be considerable care in defining terms and in planning before embarking upon a statistical study. In the simple example at hand, one should consider that heights are generally greater upon arising in the morning than before retiring in the evening and be sure that there is consistency in the time of day the data are gathered from the children. Since a child's height may change over a few weeks due to growth, all the data should be gathered within a short span of days. The method of measurement and the measuring instruments must be analyzed for repeatability and reproducibility of results. If different persons are performing the measurements, there must be training in regard to the consistent use of the measuring instruments and quality control procedures must be put in place. By focusing on the issue of operationally defining "height", most of the critical considerations will surface.

You are right to conclude that, in most statistical studies, more effort must be made to define terms operationally, to attend to measurement error, and to attend to the measurement process than to deal with sampling error. The latter can be controlled through the choice of sampling plan and sample size; with sampling error, the concerns are technical and relatively easy to handle.

Random sampling properly carried out in an enumerative study allows for the quantification of the sampling error. Deming (1954) puts it this way in his discussion of the role of the statistician in testifying about the difference between the sample estimate and the corresponding population parameter. Dr. Deming states that the statistician can testify as to the sampling error, that is, the discrepancy between the sample statistics and the corresponding population statistics if the same methodology and care were taken in processing the entire population as was done on the sample. The statistician may not be able to speak to the issue of the bias and the variability introduced by the methodology used in measuring the sample elements.

I will tell of an experience where measurement error was central in the arguments surrounding an attempt by the State of Michigan to collect over a million dollars in alleged overcharges to the Medicaid program. Michigan Department of Social Services (MDSS) has the responsibility to administer the Medicaid program in Michigan. A doctor, clinic or hospital will bill the program for services rendered to eligible persons and receive direct payment through MDSS.

Of course, MDSS has an obligation to audit these expenditures. MDSS has developed procedures for sampling the charges by a provider for a specified audit period and estimating the total overcharge. The procedures are published in Lovell (1983); this document is currently under revision. The document is made available to all medical care providers who do business under Medicaid or wish to do so.

The patients who have used the provider's services during the audit period and who had charges to the Medicaid program are the primary sampling units. The sampling procedures are stratified random sampling; the patients are stratified into strata according to the total charges under the patient's name for the audit period; for example, patients with large totals go into one stratum. (I will not discuss stratified random sampling here except to say that reasonable stratification followed by some form of random sampling from the separate strata and the appropriate combination of sample results from the separate samples is an extremely effective way to obtain precise estimates.) Lovell (1983) specifies in considerable detail the statistical procedures that are used in this enumerative application. A statistician is responsible for overseeing the implementation of the procedures for stratification and the process that selects the sample patients from each strata for the audit.

Once the sample is selected, the MDSS medical care experts will examine each payment and indicated service for each sample patient to determine the amount of overcharge. (An overcharge will be zero or even negative if it is determined that the payment for a billed service is appropriate or less than a certain amount deemed fair and reasonable.) If the MDSS experts are biased in favor of the State, then they are likely to make the judgment calls in favor of the State and against the provider, thus, tending to state larger overpayment values than would be determined by a neutral expert. If they are careless and make haphazard errors in their determinations, they will introduce additional measurement error into the sample results.

There is an opportunity for the reduction of measurement error after the initial determinations by the State. Meetings take place between the MDSS experts and the provider's experts in which persons negotiate the amount of each sample overcharge. Based on the sample values thus determined, a final estimate is made by the State of the total overcharge for the population of patients. (If there is disagreement about some of the sample determinations, each side may make its own estimate.) The State's estimate will be the total claimed liability for the provider and collection will be attempted.

Example 2.4 Removing Measurement Error in an MDSS Sample

In the case on which I consulted, the total of all the charges being audited was $1,280,050.98 coming from services for 9,324 patients. The sample consisted of simple random samples of 75 patients selected from each of three strata. A ruling by the Judge after the sample was taken had the effect of removing certain charges from the domain of the audit, which reduced the total involved to $1,141,214.82.

The ruling was that certain physicians among those whose charges were being audited under the name of the clinic in which they were employed should not have been subject to the audit. This implies that if a population patient had services from these physicians only, then the charges to that patient were not subject to audit. Most population patients had seen several physicians and the effect of the Judge's ruling was to simply reduce the figure for total charges associated with those patients. As you can see, the effect of the ruling on the total amount of the charges associated with the population patients was to reduce the total by about 11%.

An expert hired by attorneys representing the provider claimed that the methodology used by the State was inappropriate because the frame by which the sample patients were selected included the patients whose charges were not the proper domain of the audit. The Judge's ruling (after the sample had been drawn and investigated) created the situation that the expert seized upon as reason to discredit the statistical study.

The expert's claim was not based on a valid argument since the effect of the Judge's ruling was exactly the effect of removing measurement error by the assignment of the value 0 to any overcharges associated with charges stipulated by the Judge to be outside the proper domain of the audit. My testimony to this effect was supported by a third expert of great reputation; the case was ultimately settled out of hearing for an amount less than that sought in the recovery.

There is a somewhat subtle point in this case, and it apparently confused the one expert. I will illustrate the point through the following simple example of a population which has N = 15 patients. (See Table 2.1.) In this example, we use simple random sampling to estimate the total overcharge related to the services rendered to these fifteen patients. In Table 2.1, Column X gives the total charges to Medicaid for the patient and Column Y gives the amount of X that is initially determined by the State to be an overcharge. Columns X' and Y' give the results for total claims and overcharges after implementation of a Judge's ruling that certain charges were not the proper domain of the audit. Column Y" gives the overcharge figures after the removal of measurement error through negotiations and further investigations. All numbers are in units of dollars. The sample patients are indicated with a *.

For this hypothetical population of patients, the total charge subject to audit was reduced from $4,840 to $3,960 by the Judge's ruling after the sample was drawn. An unbiased estimator for the total overcharge $1,840 is provided by the estimator $N\overline{Y}"$, where N = 15 and $\overline{Y}"$ denotes the sample average for the Y" characteristic. With the indicated sample, the average is $116 so the estimate is $1,740.

We see that adjustments came after the sample was drawn. The adjustment from Y to Y' was the result of the effect of the Judge's ruling which was based on legal considerations and not on the sample values. The adjustment from Y' to Y" came with knowledge of the sample Y'-values, and, conceivably, the provider being audited may put a level of the effort to argue an individual charge depending on its type and size. A lot of the difficulty arises here because there are not available operational definitions that unambiguously define what constitutes proper medical treatment covering all of the various contingencies that may arise in the medical care of a large set of individuals. It follows that arguments can be made that question whether

Table 2.1. Estimating Total Overcharge from a Sample.

Patient	X	Y	X′	Y′	Y″
1	200	100	180	90	80
2*	350	50	200	0	0
3	440	300	400	300	300
4	250	50	200	40	20
5	220	120	210	120	110
6*	330	90	300	90	90
7	380	200	350	200	190
8*	200	80	160	70	50
9	370	220	360	210	210
10	280	100	270	100	100
11	210	110	100	10	0
12*	430	300	370	290	280
13*	480	220	400	170	160
14	500	300	460	260	250
15	200	50	0	0	0
Total	4,840	2,290	3,960	1,950	1,840

the sampling distribution based on simple random sampling (that leads to unbiasedness and other properties for the estimator) is still the appropriate distribution when such after-the-fact sample adjustments are made. However, the application of random sampling and the quantification of the precision of sample statistics would essentially be eliminated in practice if such arguments implied that the sample results were no longer useful.

Actually, provider's expert, who tried to impugn the State's sample, argued a more narrow issue and was wrong. In the final brief of the provider's attorney, we find this description of this expert's testimony:

> "The thrust of his testimony repeatedly was that one cannot draw conclusions as to a target population when the frame selected includes innumerable items which were not to be included in the projection."

(There were not "innumerable" items. Provider's attorney was blowing smoke. A good friend and attorney once told me that the strategy of some is: "if the law is on your side, argue the law; if equity is on your side, argue equity; if neither is on your side, confuse the issue." On the other hand, I was once told by a good-natured judge that an "expert", spoken "x spurt", is: "x", an unknown quantity and "spurt", a drip under pressure. I was on the witness stand and, naturally, I laughed.)

In Table 2.1, notice that Patient 15 might not have been included in the population had the State known in advance of the Judge's ruling. (Under an appeal the ruling might be reversed.) However, the inclusion of that patient does not invalidate the estimation procedure. In fact, if there is a question, it is probably prudent for the State to include any patients and related charges in the population that are possibly subject to review. The effect of having too large of a frame and population can be adjusted out after-the-fact by assigning 0's to any irrelevant elements if they appear in the sample. The greater risk comes from leaving elements out of the population that should possibly have been included. Carrying too large of an excess of irrelevant elements in a population will reduce precision for a given sample size, but it will not be reason to invalidate the statistical study.

In fact, Cochran (1977, p. 37), in writing of the method that replaces irrelevant sample values by 0's in calculating the sample estimate, states:

> "The methods of this and the preceding section also apply to surveys in which the frame used contains units that do not belong to the population as it has been defined."

I illustrated the point before the Judge using a computer simulation in which the Judge could get immediate feedback on the effect that his rulings have on the values of the population parameter and the sample estimate. The Judge appeared to have a difficult time in understanding the point.

One final comment on my role as a witness. I was qualified to give expert opinion on the sampling methodology used by Social Services in its audit. I was *not* an expert on what constitutes an overpayment, what medical tests that where performed were not necessary, what doctors had provided services that were not in the domain of the audit. . . . These were issues of a medical-legal-procedural nature for which I had no special expertise. I could testify as to the precision of the sample relative to what would come from carefully applying the same methodology to the population as was applied to the sample, but I could not testify that those results were relevant and free from biases introduced by the measurement process. Here a subject matter expert would provide the necessary testimony. ∎

In all of this, it is clear that there should ideally be an operational definition of what is an overpayment to a medical care provider. Typically, claims of overpayment center around the question of whether a particular medical test that was prescribed was, in fact, necessary given the facts concerning the patient's condition and history and around the size of the charge by the provider for the procedure. We can see that there may be room for a difference of opinion on such matters, and we expect that reasonable people may have to arrive at definitions through compromise. Of course, in a blatant case of fraud, there may be charges by the provider for services never given and the issues may be more black and white.

Agreement in advance between the State and the medical care providers as to what constitutes a fair and reasonable charge for specific procedures is an important part of limiting measurement error in this example.

Established protocols and hearing procedures to remove measurement error from the State's initial determinations of sample overpayment values are important. Proper documentation and communication of the definitions, procedures and protocols in clear terms are essential.

One of the great advantages to sampling as opposed to a complete census centers around the issue of measurement error. By concentrating resources on developing procedures to control measurement error and the application of such to the relatively small number of elements in the sample, a reasonable job can usually be done in limiting measurement error. On the other hand, the careful implementation of the procedures across the measurements for all the elements in a population may be impossible or simply prohibitive because of cost. In the Katz study of uninsured motor vehicles in Michigan, the sample size was kept relatively small in order to be sure that measurement error was essentially eliminated in processing the sample vehicles. This is very important for studies whose results are to be used in courts of law and administrative hearings. Persons wishing to impugn the accuracy of the results of a study are apt to spend considerable resources to find measurement error in an attempt to discredit the work.

Suppose that measurement error is properly addressed. There remains a subtle point to be discussed in regard to the sharing of risk stemming from the sampling error. In the experience involving the State and the medical care providers, the State wisely publishes in detail the sampling plans it uses to audit the charges from a provider. One estimator that is uses for the total overcharge for the audit period is an *unbiased* estimator. Its use implies that, if the State controls measurement error and uses this estimation technique on all of the audited providers, then the sampling errors associated with the separate estimates will average zero in the long run. Thus, though the State may be collecting more than it should from some providers and less than it should from other providers because of sampling error, the net aggregate effect of these errors is averaged out. Of course, individual providers will be feeling the effects of these individual sampling errors; some will be lucky and some not so lucky but with the medical providers being treated fairly in aggregate.

Arguably, the State takes a large enough sample for its audits to produce acceptable levels of sampling error for the individual providers. Naturally, the audited providers are individually prone to argue for the lower endpoints of the confidence interval estimates of their total overcharges. If such arguments are successful, then, on average, the State will be under-collecting what is properly due it and the taxpayers. Therefore, it is reasonable for the State to seek agreement on the terms of the potential audits and the risk sharing aspects associated with statistical estimates in advance of accepting a provider into the Medicaid system.

Communication and cooperation among various elements are essential to the overall efficiency of the medical care system as in any system.

Without these, costs are greatly increased due to legal wrangling and after the fact analysis, often in an environment filled with emotion, controversy and contention. Again, we see where efforts at securing agreement in advance as to protocols and rules and agreement in advance as to adherence to the rules can pay great dividends when working with issues surrounding a statistical study.

Section 2.5 Summary and Other Considerations

As we have seen, enumerative studies, though conceptually simple, are not so simple to carry out in a meaningful way. Much care should be extended in operationally defining terms and in deciding exactly what counts and measurements are to be taken on the sample elements.

Usually, the first question that the uninitiated person asks in designing a sample survey concerns the sample size. Statisticians know that the person responsible for the study has many things to consider before the question of sample size can be addressed. To help that person focus on the key issues, I like to tell the person to imagine that it were possible to do a complete census of the population and to imagine that there are no limitations as to the characteristics that he/she can measure. I ask the person to operationally define every characteristic that he/she wishes to measure. I ask the person to explain why these population characteristics (counts, proportions, means, standard deviations, differences, . . .) are useful for the research or decision process. I ask the person about what means he/she has to control measurement error. If the person can not properly focus on the types of characteristics he/she needs to measure or can not operationally define the terms or can not specify the role that determined population values have in his/her arguments and decisions, then I feel I can be of little help. Determining a sampling plan and a sample size are relatively simple technical matters that must follow the proper attention to the more fundamental issues.

Some researchers wrap themselves in the confusion created by sampling and countless statistical manipulations; if unburdened of the sampling considerations, they are sometimes exposed as having little capacity to use the complete census information in a reasonable way. In such cases, the researcher is not strong in the subject matter domain of the application, and a statistician can not be expected to save the study.

Simple random sampling as a sampling plan has the advantage of being extremely simple in concept and very easy to illustrate to the layperson. However, as we will perhaps better appreciate after reading Chapter 4, it has limited applicability because of the large standard deviations it induces on sample statistics and/or the high costs sometimes associated with collecting data from geographically disperse elements that it chooses for the sample. For legal proceedings there is a great value to simplicity; in the other arenas it is less important.

Stratified random sampling, cluster random sampling and systematic random sampling are devices that can be thought of as constrained simple random sampling. They constrain the possible outcomes to the sampling to sets of samples that are thought to be representative. Stratifying households in a city into both "sides of the track" and splitting the sample between the two sides is an obvious improvement over unconstrained simple random sampling of households in the city.

For a specific example, reconsider the set of male children of Section 2.4. Below you will find the ages in years and the heights in inches for the six boys.

Children:	A	B	C	D	E	F
Age:	2.1	2.4	2.3	1.9	1.8	1.9
Height:	36.2	39.0	35.3	34.9	32.6	35.6

We compare the sampling distributions of the sample average $\overline{X}$ based on two different sampling plans. One plan is the simple random sampling of n = 2 children from the N = 6. The other plan stratifies the children based on age into two strata of size three so that A, B, and C go into one stratum. Then a single child is selected at random from each stratum to produce the sample of size n = 2.

Table 2.2 gives the probability distributions for $Y = \overline{X}-33.00$ under the two sampling plans. (For convenience, we have tabled the number of outcomes out of the 15 possible under simple random sampling and the number of outcomes out of the 9 possible under the stratified sampling plan.) Note that stratified random sampling is preferred to simple random sampling here since the probability is more concentrated about the population mean height $\mu_Y = 2.6$.

Table 2.2 Sampling Distributions for Y based on the Two Plans.

.75	.95	1.1	1.3	2.1	2.25	2.45	2.55	2.75	2.8	2.9	3.95	4.15	4.3	4.6
1	1	1	1	1	1	1	1	1	1	1	1	1	1	1
1		1	1		1	1		1	1	1			1	

At the extreme of the constrained random sampling plans is judgment sampling where a person uses only his/her judgment to select the elements for the sample. In some instances for developing data for personal decision-making, it seems reasonable to do this; however, the person must be aware of the possible biases and dangers from trusting to one's own judgment in selecting the individual sample elements from the population. For matters of research and decision-making of concern to the public, random sampling plans should be used. The eminent Professor Cochran had an interesting experience in this regard. When he testified in support of a judgment sample in a hearing, he was told by the judge that the only thing that the judge had learned from his statistics course was that one should not use judgment samples.

The use of judgment in selecting samples is a vital part of the design of an analytic study. The reader should refer to Deming (1976).

I conclude this chapter, with a brief discussion of *forecasting* or *prediction*. In a forecasting problem, an estimate or prediction is made of the value of a future outcome of a process. This is quite different from estimating a value of a parameter for a fixed population. Forecasting concerns the prediction of an individual event; estimation in an enumerative study concerns the estimation of an aggregate characteristic of a fixed population.

For example, I have spent considerable effort at trying to reduce the forecasting error for yearly tart cherry production in Michigan. (Michigan leads the nation among states in the production of tart cherries.) Each year the growers and processors use a June forecast for the size of the crop to be harvested in July and August in negotiating prices and in planning for the harvest. At first consideration, the reader might think that this is simply an enumerative study since an estimate of total weight of production for a fixed population of several million trees is desired and a sample of trees and branches is taken. However, the timing requires that the sample data be gathered in June when the cherries are green and may be less than half the size of a pea, some four to six weeks before the harvest. Because the cherries are subject to drop and the cherry production is dependent on weather conditions in the intervening weeks, statistical studies are used to develop a better understanding of the drop feature of the fruit and the effects that weather conditions have on the maturation of the fruit. These efforts are essentially analytic in nature. The enumerative aspect of the model concerns the fact that a the June random sample can be thought of as being used to estimate the total count of immature cherries on the trees at that time. The analytic work provides a model that forecasts the drop and one that forecasts the increase in weight.

Forecasting is tricky business. In modeling involving data accumulated over a period of years, one must be concerned with stability and possible cyclic features. In addition, as long as weather conditions can not be forecast with precision, the cherry crop can not be forecast with precision. For example, high winds and wide spread hail storms can destroy a huge portion of the crop, and no weather forecasting model is yet capable of forecasting such events weeks in advance. In summary, even with a reasonable forecasting model you may find the forecast errors to vary fairly widely and, possibly, in an unstable manner. A model that describes the distribution of errors may be a mixture of one distribution and a distribution with some extreme values representing the effects of the rare weather events.

I once was given major responsibility in organizing a chicken barbecue. This had become an annual event in the community with a history of several years. Based on previous experience, the observed past relationship of pre-sales at various times and number of tickets purchased at the gate, and the levels of advertising and promotion, I placed an order for chicken several days before the event. I could not factor in the weather which clearly would affect the sales on the day of the event. I was anxious on the day of the barbecue; as the evening came near, I watched the last of the chicken halves and quarters disappear as the service line shortened. With over two thousand dinners served, I had called the chicken order to within a few chickens! Only a few halves and a couple of quarters remained on the grills after the last customer was served.

Later in the beer tent, I basked in the adulation of the other organizers; they thought that I had tremendous power in my statistics. I probably could have sold them a prediction for the next lotto. I probably was foolish enough to begin to believe in the power of my mind to assimilate information on a process and to forecast the next realization of the process.

The next year I tried to call it too close and suffered the embarrassment of having to refund money to many hungry patrons. I should have better assessed (guessed) the variability in the process and to have taken into account penalties for overforecasting and underforecasting.

Section 2.6   References

Cochran, William G., *Sampling Techniques, Third Edition*, Wiley, New York, 1977.

Deming, W. Edwards, *Some Theory of Sampling*, Wiley, New York, 1950.

Deming, W. Edwards, "On the Distinction Between Enumerative and Analytic Surveys," *J. Am. Statist. Assoc.*, **48**, 244–255, 1953.

Deming, W. Edwards, "On the Presentation of the Results of Sample Surveys as Legal Evidence," *J. Am. Statist. Assoc.*, **49**, 814–825, 1954.

Deming, W. Edwards, *Sample Design in Business Research*, Wiley, New York, 1960.

Deming, W. Edwards, "Principles of Professional Statistical Practice," *Ann. Math. Statist.*, **36**, 1883–1900, 1965.

Deming, W. Edwards, "On Probability As a Basis for Action," *Am Statistician*, **29**, 146–152, 1975.

Deming, W. Edwards, "On the Use of Judgment-Samples," *Rep. Stat. Appl. Res., JUSE*, **23**, 25–31, 1976.

Deming, W. Edwards, *Out of the Crisis*, Massachusetts Institute of Technology, Center for Advanced Engineering Study, Cambridge, MA, 1982, 1986.

Finkelstein, Michael O., *Quantitative Methods in the Law*, The Free Press, New York, 1978.

Gilliland, Dennis C., "Decision Rules for Canvassing Boards: The Michigan Experience," *Proceedings of the Midwest American Institute of Decision Sciences Annual Meeting*, 288–290, 1981.

Gilliland, Dennis C. and Meier, Paul, "The Probability of Reversal in a Contested Election," In DeGroot, Morris H., Fienberg, Stephen E. and Kadane, Joseph B. (Eds.), *Statistics and the Law*, Wiley, New York, 391–411, 1986.

Katz, Leo, "The Presentation of Confidence Intervals in Administrative Proceedings," *Am. Statistician*, **29**, 345–350, 1975.

Lovell, Robert G. "Medicaid Practitioner Audit System: Sample Design and Estimation Calculation," Office of Quality Assurance, *Michigan Department of Social Services*, July, 1983.

Meier, Paul, "Damned Liars and Expert Witnesses," *J. Am. Statist. Assoc.*, **81**, 269–276, 1986.

Personal Observations on Statistics and the Law

This chapter is made-up of the text of a talk that I gave to the Mid-Michigan Chapter of the American Statistical Association in East Lansing, Michigan.

Section 3.1   Introduction

In this talk, I will present some general observations concerning the statistician in the role of advisor and/or witness in legal and administrative proceedings based on personal experiences. The talk has three sections:

- What should take place at the interface of statistics and law (statistician and lawyer). Guidelines and advice for accomplishing what should take place within the present system.
- What does take place.
- Proposed rules and changes for the system that will improve the process.

The talk is based on my personal experience as a systems analyst, teacher, researcher and consultant. This experience began in 1959 when I joined Goodyear Aircraft Corporation upon graduation with the bachelor's degree. The line of experience as consultant and witness began in 1972 when three professors and I were solicited to present testimony before the State Board of Canvassers critical of the sampling procedures used by the Elections Division for a statewide petition. That led to a particularly satisfying experience since Professor Jim Stapleton and I were later asked to suggest proper sampling procedures. The procedures we suggested were adopted, later modified, and now play an important role in the Elections Division's analysis of all statewide petitions. The procedures have stood the test of time and seem now to be accepted by all parties, proponents and opponents, when it comes to statewide petition drives and the decision as to sufficiency. (Serious legal challenges have not yet been raised.)

Only a portion of consultations with attorneys and administrators actually led to appearances as a witness. I will mention those appearances and the subject matters to give you an idea of the breadth of subject matters that a statistician may be dealing with as a witness.

1. Michigan State Board of Canvassers on Petitions
2. Circuit Court—Genesee on a Contested Election
3. District Federal Court—Kent on Fraud
4. Circuit Court—Ingham on Salt Tariffs
5. Circuit Court—Wayne on Punch Card Voting
6. Circuit Court—Kent on Legal Malpractice
7. Circuit Court—Wayne on Political Gerrymandering
8. District Federal Court—Genesee on Ballot Position
9. Circuit Court—Wayne on Damages by Highway Noise

I have made additional appearances before the Michigan State Board of Canvassers, two appearances before the Michigan Public Service Commission (one as a witness critical of a study that was commissioned by a utility and the other to report the results of a statewide survey of utility poles that I carried out for the cable television industry), an appearance before the Insurance Bureau, and, finally, an appearance before an administrative law judge hearing a dispute between the Michigan Department of Social Services and a medical care provider concerning alleged overcharges. In addition, I have prepared written statistical material for briefs presented to the U.S. Supreme Court on matters related to the annexation of townships by cities in Michigan.

I appreciate the opportunity that preparing for this talk has given me to collect my thoughts and observations based on these experiences and for having the chance to share them with you. At the end of the talk, I will provide you with references and a list of recommendations coming from the Panel on Statistical Assessments as Evidence in the Courts and reported in Fienberg (1989). There will be time for discussion and comments.

I will not be discussing deep questions of what type or mode of statistical and probabilistic thinking is appropriate in the arena of hearings and judicial decision-making. My experience has been that efforts go at (a) understanding the uncertainty that has lead the attorney to solicit my advice; (b) finding a way to explain this uncertainty (this may require that a statistical study be performed); (c) reducing complex issues related to uncertainty to simpler questions about models; (d) examining the assumptions of models in regard to their empirical justifications and the robustness of conclusions based on the models given departures from the assumptions; and, finally, (e) preparing written testimony, exhibits and questions and answers for direct examination.

Questions of Bayesian/Frequentist do not seem to emerge; rather the details and structure of the problem being faced seem to dictate a most reasonable approach to me. I am not prepared to deal with philosophical questions in the context of particular legal cases. These questions have not been answered to the satisfaction of the statistical profession and philosophers dealing with the question of inductive logic. I believe that no one general philosophy or approach to probability will be satisfactory in all cases. When there is randomization induced by an experimenter in a sample survey or controlled experiment, the frequentist approach seems useful for framing the induced uncertainty. In one of my papers on a contested mayoral election I used a model with a specified but data-neutral parameter value to quantify the uncertainty in the outcome. Later, based on additional data, I came to realize that it would be more appropriate to assess the likelihoods by mixing the model with a probability distribution over the parameter I had earlier specified. Here I used a data-based "prior" distribution to model uncertainty in a parameter. I paraphrase William Hunter in saying that data are the basis for objective action and, presumably, for opinions, and the witness should be prepared to explain the basis for his/her opinions.

I would be remiss not to mention an error that seems to be often made by witnesses and persons using statistical reasoning. This is the error of describing the significance of a particular test statistic relative to a given null hypothesis as the probability that the null hypothesis is true. In regard to petitions, let p_0 be the proportion of valid signatures necessary for sufficiency and let $H_0: p \geq p_0$. If the sample proportion p has significance .035, then the error would lead to the statement that the probability that the petitions have a sufficient number of signatures is .035. We have seen this interpretation of significance given in this context by a technical advisor to the Elections Division. Fienberg (1989) cites examples where this interpretation of significance probability has been made in court.

Section 3.2   Guidelines and Advice to Help Get the Facts Presented for Decision-Makers in the Present System

1. *Judges, decision-makers and attorneys should be trained to some extent in statistics,* at least to the point of understanding reasonably well such concepts as sampling error, measurement error, bias, variability, stability over time, . . . The citizenry making up juries are the decision-makers in some cases.

2. *The statistician should not overstate the power of statistical theory;* it is a theory after all and seldom does it apply with the degree of certainty

that makes, for example, a 95% CI more than an approximate or nominal 95% CI.

Opinions should not be given at hearing in a flip and cavalier manner. The same concern must hold for opinions offered in consultations with attorneys and administrators as legal theories are being developed. The statistician may have an undo influence on the legal theory that evolves. I have found that too often attorneys are selective in their hearing and that selections from wide ranging discussions will often lead to overly optimistic assessments of the importance and/or objectivity of statistical analyses. With a statistician with little understanding of law and a lawyer with little understanding of statistics, you may have the blind leading the blind unless great care is exercised.

3. *The statistician should clearly understand and state the limits of his/her expertise.* Unless the statistician is also a subject matter expert in the domain of the case, he/she should restrict opinions to issues such as sampling error, interpretation of regression coefficients, presentation of data, methods of sampling, bounds on bias, . . . I had an experience before the Insurance Bureau where, based on an almost spontaneous decision, I testified on risk. I did not help my side in that case but learned how little I knew about risk theory in the jargon of the insurance business.

Deming (1954) puts it this way in regard to sampling. The statistician can testify as to the sampling error, that is, the discrepancy between the sample statistics and the corresponding population statistics if the same methodology and care were taken in processing the entire population as was done on the sample. The statistician *can not* testify as to the importance and relevance of the population statistics to the case and may not be able to speak to the issue of bias introduced by the methodology.

Example 3.1

When Professor Jim Stapleton and I proposed sampling procedures for the Elections Division in 1973, we used the term "valid" to describe a signature that had passed a certain number of tests for validity required by law and administrative procedure. We could testify as to the precision of the sample proportion of "valid" signatures in so far as it estimates the corresponding population proportion. We recognized that the tests did not test for duplication and so stated this. Absence of knowledge of the magnitude of duplication rates at that time prevented us from making any kind of statement on the usefulness of the sample estimates, but the procedures were adopted absent that knowledge. In the meantime, we have learned more about signature duplication rates that come from statewide petition drives. ■

It is important for the statistician to clearly communicate the limits of his/her expertise so that the attorney can solicit other experts in a timely fashion and otherwise prepare the case.

4. *The statistician should remain neutral and unbiased and not be selective in regard to the statistical information that is reported.* This is an arguable position given the statistician's obligation to the attorney and his client. I recommend that any statistician with the neutral stance should so inform the attorney since the attorney must operate in an adversarial arena. It is difficult to remain neutral and not take sides. Meier (1986) and Deming (1954, 1965, 1986) make excellent points here.

I like the quote from Shepard found in Meier, p. 273 , since it suggests a contrasting position and is based on an attorney's perspective. From Meier we have

> John C. Shepherd of St. Louis, a distinguished trial lawyer, who was President of the American Bar Association in 1984–1985, spoke to a conference for lawyers on relations with the expert witnesses (Shepherd 1973), and this is what he said:
>
> "Many people are convinced that the expert who really persuades a jury is the independent, objective, nonarticulate type . . . I disagree. I would go into a lawsuit with an objective, uncommitted, independent expert about as willingly as I would occupy a foxhole with a couple of non-combatant soldiers.
>
> If you find the expert you choose is independent and not firmly committed to the theory of the case, be cautious about putting him on the stand. You cannot be sure of his answers on cross-examination. When I put an expert on the stand, he is going to know which side we are on.
>
> The trial lawyer must make of the expert a convincing, persuasive witness. The lawyer deals in words, and he knows how to put the package together to impress the jury favorably. It is his job to instruct the expert, an exercise requiring great tact and firm conviction. (pp. 21–22)."

5. *The statistician should have written opinion prepared enough in advance* so that there is time for the other side to develop counter theories, perform studies, and depose the statistician. (See The Manual for Complex Litigation (1985).) By preparing written testimony, the statistician will sharpen his/her own thinking and will better understand the limits of his/her own knowledge and opinions.

6. *A statistician who is directing a study should keep a journal of activities and decisions just as any good scientist would.* It is fairer to all concerned to have such a record and the benefit to the credibility of the study counters the fact that the same journal will invariably expose some weaknesses of the methodology in its application.

I conducted a sample survey of the population of utility poles in Michigan to which were attached cable television lines as of a certain date. I was conducting the study for the cable television industry that rented space on poles for their attachments. There was no single frame for the poles in the population, and I had to work with many local television companies in developing the sample. There were different ways of developing sampling frames for the different local companies because of the variation in the records that were kept by these companies. I believe I was successful in limiting bias due to nonequal probability sampling that necessarily resulted from the use of imperfect frames and the various models of selecting poles. As a scientist, I kept very detailed records of all plans, communications and directives, and I made on-site inspection of the work of local companies and the survey crew. I kept records while realizing that these same records would provide ammunition for the experts hired by the utilities that owned the poles. After cross-examination, I remained pleased with my decision to keep detailed records.

This reminds me of what I learned from a telephone call to a so-called expert statistician. I had read testimony from this person offered before the Ohio Public Service Commission in which an estimate and standard error were presented based on a sample survey directed by the individual. The testimony did allow me to determine that the sample units were selected by cluster sampling and this was confirmed in the conversation. The smallness of the standard error suggested to me that the statistician had used the formula for standard error appropriate for simple random sampling and inappropriate for cluster sampling. The telephone conversation confirmed this. When I asked to be sent the raw data, I was told that it had been destroyed very soon after being generated (thus ending an audit trail that would lead to measures on intracluster correlation and a correct estimate of sampling error). I was very depressed by this behavior by someone who gave testimony as a statistician. I judge him to be more unethical than incompetent, but I am not sure about that.

I hope that you as statisticians will maintain your self-dignity, will walk the high road, will remain true to the code of ethics of a scientist. I believe that this will enhance the chances for facts to be fairly presented and argued in hearing or court.

7. *There should be no surprises for your attorney and client.* Let them know at the outset your ethical standards and position in regard to advocacy. Help prepare questions and be sure that they understand the statistical assessments and analysis to the extent possible.

Section 3.3 What Does Take Place

1. *Adversarial statisticians sometimes argue their interpretations; in some cases they simply stonewall with "that is my opinion" with little demonstrated basis in fact for the opinion.* It seems fair that any overly precise statement should be backed-up by equally precise and unambiguous facts and attention to detail. If someone attaches an exact confidence level or significance level to some statistical procedure, that someone should be prepared to back-up the statement.

2. *Too many persons are accepted as experts who should not be accepted.* A Ph.D. alone is not a basis for qualification. Statistics and its ap-

plication form a broad spectrum of ideas and experiences, and a Ph.D. statistician may not be qualified in certain areas. Some very misinformed persons take the stand and clutter the records with inappropriate and misleading statements.

3. *The question and answer format for the presentation of evidence gets in the way of presenting cogent, coherent analyses and descriptions.* However, my experience has been that judges are very understanding in this regard and allow the expert much latitude in his/her presentations.

4. *The judge or decision-maker will make a ruling, sometimes after much delay, without the help of a court appointed expert or clerk who might have a background to understand the statistical issues.* The natural tendency for the decision-maker is to duck the technical issues and decide on the basis of other, more familiar, grounds.

5. *There is a loss of respect for the statistical profession on the basis of what currently takes place.* Meier (1986, p. 272) quotes the British Medical Journal (1863) to show the loss of respect for medical evidence in some quarters over 100 years ago.

Section 3.4 Changes in the System to Improve the Process

I do not have time to list specifics based on my personal observations. Please read Deming (1965) and Meier (1986). I also like the ideas of Finkelstein (1978, Chapter 7) in regard to protocols. More protocol should be suggested and accepted. The Manual for Complex Litigation (1985) speaks out on timelessness, discovery, admissibility. If the courts would promote protocols and rules that will expedite debate and eventual understanding in regard to technical matters, the present system will function reasonably well. The power that a court appointed expert has is a real concern although I believe such experts are necessary in some cases. (See Coulam and Fienberg (1986).)

Section 3.5 References

Ad Hoc Committee on Professional Ethics, "Ethical Guidelines for Statistical Practice," *Am. Statistician,* **37**, 1–20, 1983.

Bross, Irwin, D.J., "When in Washington, Tell the Truth, the Whole Truth, and Nothing but the Truth, and Do So Intelligibly," *Am. Statistician,* **34**, 34–38, 1980.

Coulam, Robert F. and Fienberg, Stephen E., "The Use of Court-Appointed Experts: A Case Study," In DeGroot, Morris H., Fienberg, Stephen E. and Kadane, Joseph B. (Eds.), *Statistics and the Law,* Wiley, New York, 305–332, 1986.

DeGroot, Morris H., Fienberg, Stephen E. and Kadane, Joseph B. (Eds.), *Statistics and the Law,* Wiley, New York, 1986.

Deming, W. Edwards, "On the Presentation of the Results of Sample Surveys as Legal Evidence," *J. Am. Statist. Assoc.,* **49**, 814–825, 1954.

Deming, W. Edwards, "Principles of Professional Statistical Practice," *Ann. Math. Statist.*, **36**, 1883–1900, 1965.

Deming, W. Edwards, "Principles of Professional Statistical Practice," In Kotz, Samuel and Johnson, N. L. (Eds.), *Encyclopedia of Statistical Sciences*, **7**, Wiley, New York, 184–193, 1986.

Fairley, William B. and Mostellor, Federick (Eds.), *Statistics and Public Policy*, Addison-Wesley, Reading, Mass, 1977.

Fienberg, Stephen E. (Ed.), *The Evolving Role of Statistical Assessments as Evidence in the Courts*, Springer-Verlag, New York, 1989.

Finkelstein, Michael O., *Quantitative Methods in the Law*, The Free Press, New York, 1978.

Gibbons, Jean D., "A Question of Ethics," *Am. Statistician*, **27**, 72–76, 1973.

Manual for Complex Litigation, Second Edition, Federal Judicial Center, Washington, D.C., 1985.

Martin, James A., "The Proposed Science Court," *Michigan Law Review*, **75**, 1058–1091, 1977.

Meier, Paul, "Damned Liars and Expert Witnesses," *J. Am. Statist. Assoc.*, **81**, 269–276, 1986.

Moensseus, Andie A., Moses, Ray Edward, and Inbau, Fed E., *Scientific Evidence in Criminal Cases*, The Foundation Press, Mineola, New York, 1973.

Piehler, Henry R., Twerski, Aaron D., Weinstein, Alvin S. and Donaher, William A., "Product Liability and the Technical Expert," *Science*, **186**, 1089–1093, 1974.

Van Matre, Joseph G. and Clark, William N., "The Statistician as Expert Witness," *Am. Statistician*, **30**, 2–5, 1976.

Zeisel, Hans, "Statistics as Legal Evidence," In Kruskal, William H. and Tanur, Judith M. (Eds.), *International Encyclopedia of Statistics*, **2**, The Free Press, New York, 1118–1122, 1978.

4

Probability Models

In this chapter, I introduce some elementary notions of probability and some basic families of probability models; for each family, I give some experiences in which the models played a role. These families are parametric families of probability distributions called Binomial, Hypergeometric, Poisson and Normal (Gaussian). Properties of these families are covered in detail in most every book in introductory probability and statistics.

I state the probability functions for each of the three families of discrete distributions and the probability density functions for the normal distributions. We will often refer to these functions simply as probability distributions rather than as probability functions and probability density functions. The mean and standard deviation are given for each distribution. Unless otherwise noted, the non-trivial calculations that I report involving Binomial, Poisson and Normal distributions were performed with programs written for the PC by Fabian (1986).

Section 4.1   Some Elementary Probability

In Rao (1989, p. 59), we find the quote from Henry Theil: "Models are to be used but not to be believed." A probability model for the outcomes of an experiment or a phenomenon that exhibits variability is no exception. Such a model is useful if it predicts, within reasonable limits determined by the model itself, the outcomes or properties of aggregates of outcomes. Such a model can not be proved or demonstrated to be true so we will never have reason to believe that a model holds exactly.

For example, the model which specifies that the number of H's in 10 tosses of a coin has the Binomial distribution with n = 10 and p = .5 is a useful model for understanding and modeling the variability in the number of H's across repetitions of the experiment. However, the model can not be believed; the assumptions that the trials are independent and the coin is fair (p = .5) can not be proved to hold for this experiment. Statistical methods exist which can be used to estimate the lack of independence and to estimate p if this parameter is not specified in the model. These are discussed in standard texts in statistics.

Consider an experiment and a random variable X with four possible outcomes {0, 1, 2, 3}. A model for the probabilities concerning X is a specification of probabilities for the outcomes or values of X. For example, suppose the model specifies

x	0	1	2	3
p(x)	.4	.3	.2	.1

Here the mean or expected value of X is $\mu_X = 1$ and the standard deviation of X is $\sigma_X = 1$. If the model is good, then, in a large number of independent repetitions of the experiment, the relative frequency of occurrences of the outcome X = 0 will be about 40%, and, for all the outcomes of X, the average value of X will be about 1 and the standard deviation will be about 1. The model may accurately predict aggregate behavior of the set of outcomes without accurately predicting the outcomes for the individual trials within the experiment.

By Chebyshev, for any probability distribution, the interval $\mu \pm k\sigma$ contains at least $(1 - 1/k^2)100\%$ of the probability. For a normal or approximately normal distribution, the interval $\mu \pm \sigma$ has about 68% of the probability, the interval $\mu \pm 2\sigma$ has about 95% of the probability, and the interval $\mu \pm 3\sigma$ has over 99% of the probability. In general, the smaller the standard deviation, the more concentrated is the probability distribution about the mean of the distribution.

All probability is essentially *conditional probability*, that is, based on some model or data set. A model may be reasonable in that its applicability for the situation being modeled is based on evidence or reasons such as symmetry. Therefore, the choice of model depends upon evidence in the form of data or deductions so that the calculated probabilities are conditional on that data or deductions. The more astute the person who selects a model for a particular situation, the more credible the model. The model is not to be believed, but it may be useful and provide for a good description of variability.

As a young person I read that, before the first atomic bomb was exploded at Los Alamos, a calculation was made that assessed the chance of the explosion unleashing a chain reaction that would destroy the earth was some 1 in 1,000,000. (I do not remember the probability exactly only that it was small.) I was struck by the fact that such a number being model-based is not to be trusted to be exact in any sense since it did not or could not depend on the unknowable (the theory, facts and evidence unknown at the time and unknowable at any time). The 1 in 1,000,000 may have been useful and somewhat credible up to several orders of magnitude, but it was not to be believed.

Understanding the basis for the probabilities in a model can be extremely important as we will see from the next two examples.

Example 4.1 Simpson's Paradox

Based on empirical observation, radical mastectomy may be conditionally more effective than partial mastectomy in the treatment of each grade of a breast cancer, yet its "overall" effectiveness may be less. This apparent contradiction is an example of what is called *Simpson's Paradox*. Simple tree diagrams can resolve the paradox. For simplicity's sake, suppose that there are three grades of involvement of the disease. Figure 4.1 shows *hypothetical* rates of Success S and Failure F for each treatment as applied to large groups of patients. These rates were chosen to illustrate the basis for Simpson's paradox. We see that, for each grade of the breast cancer, the radical mastectomy has higher probability of S than has the partial mastectomy. However, the overall rate of S for the radical mastectomy is $[.3(.05)(.9) + .4(.5)(.8) + .3(.95)(.4)]/[.3(.05) + .4(.5) + .3(.95)] = .575$ and the overall rate of S for partial mastectomy is $[.3(.95)(.8) + .4(.5)(.7) + .3(.05)(.3)]/[.3(.95) + .4(.5) + .3(.05)] = .745$.

The explanation is apparent; the radical surgery has been used more extensively on the advanced Grade III of cancer while the partial mastectomy has been used more extensively on the less advanced Grade I of cancer. The conditional probabilities showing radical mastectomy to be more effective conditional on grade, for all grades, is the basis for choosing radi-

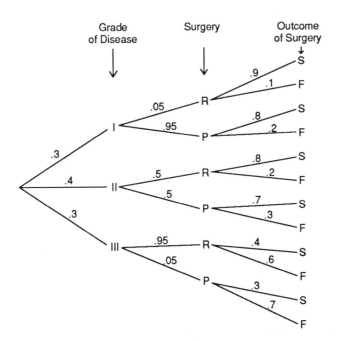

Figure 4.1 Tree Diagram for Results of Surgery

cal over partial mastectomy; someone ignorant of the explanation for Simpson's paradox might choose partial over radical mastectomy based on its overall rate of success. Bickel, et. al. (1977) discusses Simpson's paradox in connection with rates of graduate admissions to Berkeley for males and females. ∎

Example 4.2 A Prisoner's Dilemma

I will now discuss a version of the *Prisoner's Dilemma* as an example of another apparent paradox that is resolved by the careful use of probability. Two of the Prisoners {A, B, C} have been selected at random to die the following morning, and they are to be kept uninformed as to the outcome of the draw that evening. A asks the Warden to give him the name of one of the prisoners {B, C} who is to die, arguing that one of these individuals is to die and that revealing the information will not shed any light on his (A's) outcome. However, after being told that B is to die, A is relieved to know that he is now one of two prisoners {A, C} who is to die and "therefore" has only a 1 in 2 chance of death. Receiving what appears to be noninformative information has allowed A to cut his probability for death from 2 out of 3 to 1 out of 2. We will resolve this apparent contradiction with a use of a tree diagram. A probability model for this Prisoner's Dilemma is found in Figure 4.2.

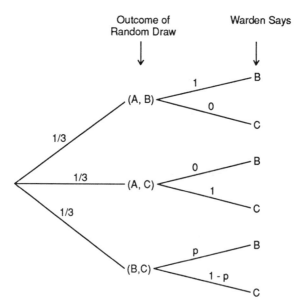

Figure 4.2 Tree Diagram for Prisoner's Dilemma

We see that the conditional probability that A is to die given the information from the Warden that B is to die is $[(1/3)(1) + (1/3)(0)]/[(1/3)(1) + (1/3)(0) + (1/3)p] = 1/(1 + p)$, where p is the probability that the Warden will name B when in fact both B and C are condemned. If $p = 1/2$, that is, the Warden decides between naming B and C by the flip of a fair coin when both are condemned, then the conditional probability is 2/3, the same as the probability that A was chosen to die with the random sampling. However, suppose the Warden dislikes A and knows of his friendship with B. Being sadistic in nature, the Warden is prone to reveal the name B if B and C have been chosen to die. Knowing this, A might model the situation with $p = 1$, in which case, if he hears "B" from the Warden, then his chance of having been condemned to die is 1/2. (Under this specification for p, if A hears "C" from the Warden, he has learned that he is condemned to die.) ∎

I convey an experience where a tree diagram was used to save a student some money. In 1980, I received a call from a graduate student in business who was about to embark on a trip to the Bahamas to gamble at the craps table. He claimed to have a betting strategy which gave him an expected payoff that was positive. It was a $10 bet on "Don't Pass" followed by a place (hedge) bet with the roller if he came out with certain points; specifically, betting $12 on the point if it was 6 or 8 and betting $10 on the point if it was 4, 5, 9 or 10. I used a tree diagram and knowledge of the probability of such outcomes as rolling a 4 before a 7 in a series of independent rolls of a pair of fair dice to model the situation and to find the flaw in his analysis. With the rules and payoffs for place bets as he described them, I determined that his strategy had an expected payoff of –$0.39. His analysis did not properly use the tree diagram and the multiplication rule that combines the conditional probabilities.

Example 4.3   Jai-Lai: Covering all Possibilities
In the March 5, 1983 Miami Herald, Andrew Beyer reported an interesting bet at Palm Beach Jai-Lai with $524,288 wagered to cover all possible combinations of winners of the Pick Six gamble. With eight teams competing in each of the six contests, a total of $8^6 = 262,144$ winning combinations were possible. Here $2 was placed on each of the outcomes guaranteeing a win of the jackpot that had grown to $551,331 through carryover money from 146 sessions in which no one had called all six winners. Since the winner would get back most the money wagered, the bet would be very profitable as long as nobody else picked six winners. Here the gambler had done his homework. Palm Beach Jai-Lai generally had small crowds and, supposedly, jai-lai does not lend itself to intelligent handicapping making the outcome as difficult to call as if it were generated at random. The bet was made that evening after an advance call to alert Palm Beach Jai-Lai that the large wager was to be made. The result was the largest

payoff $464,038.20 = $988,326.20 − $524,288 in the history of American parimutuel betting. You see, it can pay to learn about counting rules. ■

Example 4.4   A Baseball "Statistician"

A Barry Rohan column in the Detroit Free Press in April, 1983 credits a onetime psychology graduate student turned expert in the statistics of baseball as being the first to formulate baseball's Whirlpool Principle, which postulates that very good teams and very bad teams are propelled towards mediocrity. The expert is quoted as saying "about 70% of the teams that improve in one year decline in the next, and, conversely, that about 70% of the teams that decline in one year improve in the next". He goes on to say "that he believes that this is simply because losing teams are willing to make changes and winning teams are not". Professional statisticians are well-aware of the general principle of regression toward the mean and can explain such phenomena as a natural consequence of random variation. If three numbers X_1, X_2, X_3 are drawn at random from a given continuous distribution, then the conditional probability that $X_3 < X_2$ given that $X_2 > X_1$ is 2/3 and the conditional probability that $X_3 > X_2$ given that $X_2 < X_1$ is 2/3.

The point is that an explanation for the 70% figures involving baseball teams' performances is provided through an understanding of random variation in a stable system. As many persons are prone to do, the baseball "expert" offers a deterministic explanation to explain variation. I believe that an underlying aspect of Deming's thinking is a clear appreciation that variation is often best explained through common causes rather than special causes and pet theories. As we saw in Chapter 1, Deming uses the Bead Experiment to make the point that managers and supervisors attribute variation to their workers that is due to a system and not special causes. ■

Exercise 4.1. You can easily illustrate the phenomenon discussed in Example 4.4 by rolling a die until three distinct numbers have been observed. Record the distinct digits in the order they were observed and denote them by X_1, X_2 and X_3. Run this experiment 30 times. Record the number of times u that $X_3 < X_2$ among the v trials where $X_2 > X_1$ and compare the ratio u/v with the probability 2/3. Record the number of times w that $X_3 > X_2$ among the x trials where $X_2 < X_1$ and compare the ratio w/x with the probability 2/3. ■

Section 4.2   Binomial Distributions

The Binomial probability distributions have two parameters: n = # of trials and p = probability of success on a single trial. For a Binomial random variable X, the probability distribution, mean and standard deviation are

(4.1) $p(x) = C^n_x \, p^x \, (1-p)^{n-x}, \; x = 0, 1, \ldots, n$

and

(4.2) $\mu_X = np, \qquad \sigma_X = \sqrt{np(1-p)}.$

We denote this distribution by B(n,p).

Exercise 4.2. (a) Derive the probability distribution for X = # of H's in n = 3 independent tosses of a fair coin. Use the tree diagram found in Figure 2.1. Compare the results of the derivation with the values given by (4.1). (b) Perform the experiment of tossing a coin n = 3 times a total of 32 times. Record the number X of H's in each experiment. (c) Calculate the relative frequency of occurrences of the outcome X = 3 and calculate the average and the standard deviation of all 32 outcomes for X. Compare these empirical results with what is predicted by the model.   ■

If the model is a reasonably good model for the experiment, the empirical results should correspond to the predicted results within the range of chance variation determined by the model. Based on the B(3,.5) model, the number Y of occurrences of 3 H's across the 32 trials in Exercise 4.1(b) has distribution B(32,1/8). We see that the expected value of Y is 4 and its standard deviation is 1.87. I have calculated using this distribution that P(1 ≤ Y ≤ 7) = .9466. Even though the readers will be working with different coins as they do Exercise 4.2, I expect that most will observe 3 H's in the 32 trials between 1 and 7 times, inclusive.

Example 4.5   The Spock Case

Empirical results that are extremely out-of-line with what is predicted by the model are evidence that the model is not a reasonable model for the phenomenon being modeled. For example, a law clerk assigned to the court hearing the charges against Dr. Benjamin *Spock*, the well-known writer and pediatrician, was supposed to have selected jurors at "random" from a list of citizens eligible for jury duty in the Boston area. (See Zeisel (1969).) The list had about 56% women, yet, the 100 selected had only 9 women. With the B(100,.56) model, the number X of women selected has expectation 56, standard deviation 4.96 and $P(X \le 9) < 10^{-22}$ The probability $P(X \le 9)$ is called the *statistical significance of the observation 9 in the direction of too few women*. The observation x = 9 is overwhelming evidence against the B(100,.56) model, and, therefore, against the hypothesis that the clerk had indeed used simple random sampling in selecting the pool of 100 potential jurors from the large list of citizens.   ■

Example 4.6   Hazelwood

(See Meier et.al. (1986).) There is an important distinction between the use of statistical significance in the Spock proceedings and that in the discrimination case against the *Hazelwood* School District. Hazelwood is in the suburban St. Louis area. In a period of rapid expansion from 1972 to 1974, it apparently hired 405 teachers of whom 15 were minority. In legal proceedings concerning alleged discrimination in hiring, it was argued that the labor pool of qualified teachers was 5.7% minority (with St. Louis excluded). The Binomial model B(405,.057) was used in an attempt to quantify the discrepancy between the observed 15 and an expected 23.1 hires of minorities under random selection from this labor pool. The model-based probability of observing 15 or fewer is the probability $P(X \leq 15) = .0456$. This is the statistical significance of the observation 15 in the direction of too few minorities. This is marginal evidence *against the model B(405,.057)* but *not direct evidence in a proof of discrimination* since there is no statutory requirement that hiring be made by random draw. ■

In Spock, the model arguably captured the intent of the procedural requirements for selecting citizens for the jury pool; in Hazelwood, the model served as a benchmark at best. In discrimination issues, a Binomial model-based statistical significance might be used to raise a flag and perhaps to justify a shift of the burden of proof or explanation to the defendant for the under-representation of a protected class.

Example 4.7 The Michigan Three Digit Lottery

The Michigan Lottery generates a three digit number each day of the week except for Sundays and certain holidays. It uses a mechanism which supposedly simulates independent random selections from {0, 1, 2, . . . ,9} for selecting the three digits. With this model, each of the 1000 three digit numbers in the set S = {000, 001, 002, . . . 999} is assigned probability .001. If I were to play the Michigan Lottery Three Digit Game with a straight bet of $10 on the outcome 123, I would be accepting a gamble wherein I would risk losing $10 with probability .999 for the opportunity to win $5,000 with probability .001. (Here I have used the probability model that, I believe, best describes the process by which the three digits are generated.) The real implications for this gamble and repeated acceptance of this gamble are considerable. If I desperately need $5,000 and this is the only game in town, it is reasonable for me to accept this gamble.

Long term play will have serious consequences for me. The mean or expected payoff (profit) to me per play is –$5 and its standard deviation is $158.03. Across 1000 plays (about three years of play), I can expect to receive a total payoff of –$5,000 (winning once and losing 999 times). The following is a table of the probability distribution for the total payoff to me across 1,000 straight bet gambles of $10.

Payoff($) y	–10,000	–5,000	0	5,000	10,000	15,000	....
Prob p(y)	.368	.368	.184	.061	.015	.003	....

Here the payoff random variable is $Y = 5,000X - 10,000$, where X has the Binomial Distribution $B(1000,.001)$. From this fact,

$$\mu_Y = 5,000 \ \mu_X - 10,000 = -5,000$$

and

$$\sigma_Y = 5,000 \ \sigma_X = 4,997.50.$$

The Three Digit Game has available a bet called a box bet. If I were to bet $10 on 123 "boxed", I would win $5,000/6 if any of the six permutations of 123 were the outcome. With one of these gambles on each of 1,000 days, the probability distribution for the total payoff across the 1,000 is different from the one reported above for the straight bet. The distribution will reflect the more conservative nature of the box bet; it has the same mean –5,000 and a smaller standard deviation 2,035.11

Covering every three digit number on a single day with a straight bet of $10 is exceedingly foolish. Here the expected payoff is also –5,000 and the standard deviation is 0; it is certain that the outcome will be a loss of $5,000.   ■

A former colleague made an effort at detecting imbalance in roulette wheels in Las Vegas and reports his experience and some theory that it motivated in Ethier (1979). The gaming houses are very skilled at administering mechanisms that are intended to produce outcomes following specified probability distributions. Through rotations of devices, sufficient data can not be gathered on any one device to detect imbalance and favorable betting situations.

Example 4.8   The Three Digit Lottery Revisited

In 1980, I received a call from a retired FBI agent acquaintance who was then working with the Michigan Attorney General's Office. He was calling to get an opinion in regard to issues raised in a letter by a citizen of the State in which it was pointed out that the Three Digit Game was producing too many outcomes in which digits were being repeated. It is fairly easy using a tree diagram for the process of generating the three digits to show that the model-based probability is .280 for an outcome of the type where a digit is repeated, e.g., 000, 001, 121, 988. The citizen pointed out that, for the 132 experiences between May 2, 1980 and October 2, 1980, inclusive, three digit numbers with repeat digits had occurred 52 times. This is an excessive rate of 39.4%. Using the model $B(132,.28)$, the probability of observing 52 or more is .00308. This is the statistical significance of the observation 52 in the direction of too many three digit numbers with repeat digits.

This was cause for concern. However, the significance should be discounted to some extent in this case since the sample of days that was examined and reported was self-selected by the citizen. In other words, he may have purposely ignored earlier data that showed rates for such three digit numbers that were nearer to what would be expected. I researched the earlier record, and it showed that, for the 132 experiences from November 29, 1979 through May 1, 1980, inclusive, three digit numbers with repeat digits had occurred 36 times, a rate of 27.3%. The 68 experiences from October 3, 1980 to December 20, 1980, inclusive, contained 22 outcomes with repeat digits, a rate of 32.4%. All 332 experiences combined had 110 outcome with repeat digits, a rate of 33.1%. The statistical significance of this last result based on the model B(332,.28) is $P(X \geq 110) = .0229$ in the direction of too many three digit numbers with repeat digits. (Following this example, I will use a series of simulated coin tosses to illustrate the effect of self-selection.)

I was concerned based on the above data and an unfortunate experience in Pennsylvania that suggested that criminals were aware of the potential profit from the manipulation of the drawing. According to newspaper accounts, there had recently been a breach of security surrounding a similar game in Pennsylvania that allowed for a manipulation of the mechanism for generating the digits. For each of the three devices for generating the three digits, the pingpong balls carrying the digits other than 4 and 6 were weighted down with water injected with a hypodermic needle. This made it virtually certain that the three digit number would involve only 4's and 6's. The criminals covered all 8 such three digit numbers with lots of money; if $1,000 were bet on each, the payoff would have been $500,000 minus the $8,000 investment. (I believe that the outcome of the rigged lottery was the infamous "666".)

The criminals were discovered through a routine computer search of the betting and the fact that many persons with the same surnames were winners and betting heavily on the 8 outcomes that would result from the manipulation. A subsequent viewing of the videotape of the drawing showed that the pingpong balls were not behaving as expected. The criminals were caught.

I questioned my acquaintance in the Attorney General's Office about the security that surrounded the Michigan system and was told of a very elaborate security, testing and verification system. I warned him of the fact that even small departures from equal probability for the digits will create favorable bets and that such manipulation when rotated can not be detected with statistics. Since the Lottery is in business to make money for the State, a lottery must be administered with care and intelligence.

You may find it surprising to learn that if 3 digits, for example, 1, 2 and 3 are each given probability of selection .15, and each of the other 7 digits are given probability .07857 in determining each of the three digits going into a three digit number, then the chance of getting one of the 27 three digit numbers involving 1, 2 and 3 is .091125, and a bet on any or all of these numbers has positive expected payoff. As another example, if the digit 1 has probability .126 on each determination of a digit, the outcome 111 has probability $(.126)^3 > .002$, and the expected payoff from a bet on 111 is small but positive. (It is possible to prove that any departure from equal probability for the individual digits makes the probability of a three digit number with repeat digits greater than .28.) ∎

Exercise 4.3. Based on the model in the last paragraph of Example 4.3, show that probability is .091125 for an outcome where all three digits are from {1,2,3}. ∎

Often persons use results they observe to build models for the process that generated the observation. This is quite appropriate and part of the cycle of model-building and testing. However, people are biased observers, and, in judging the real significance of observations, we must be wary of significance calculated from data that were observed prior to the formulation of the model, that is, not specifically generated to test the model. In Example 4.4, we discounted the model-based calculation of the significance of the citizen having observed 52 outcomes of three digit numbers in 132 consecutive generations because the report ignores other relevant experience with the lottery numbers.

To illustrate the concept in a simple way, consider the string of 200 outcomes from 200 independent tosses of a fair coin. (I used a the programming language Microsoft QuickBASIC 4.0 to run a computer simulation of the experiment rather than toss a coin. I generated 200 random numbers U from (0,1) and took the outcome U > .5 to be H's.) I ran the simulation just once and report the results in Table 4.1. By reading from left to right and down you have the sequence in the order of its generation.

Table 4.1. Result of Two Hundred Tosses of a Fair Coin.

TTTHTTTHHT	TTTHHTHTTH	HTHTTTTHTH	THTHTTTHTH	HTTTTHTTTH
TTHHHTTHTH	HHHHHHHTHH	HHHHTTTTHT	HHTHHTTTTH	HHHHHHTTTT
TTTHTTTTTH	TTTHTTHTHH	HHTTHTTTTT	THTTTHHHTH	TTTHTHTHHH
HHTHTHHHHH	HTHHHTTHTT	HHHHTTTHHH	THHTHTTHHH	HHHTTTTTHH

If a fair coin is given 10 independent tosses, the chance for 9 or more H's is .0107. However, note that there are many subsequences of 10 consecutive trials among the 200 trials in which H's occurs 9 or more times.

One *can not* regard the result of 10 consecutive outcomes where the subsequence was selected because of its number of H's as the result of a random sample of size 10 from the coin since this selection process reflects a mechanism the observer has introduced. Here if the observer were biased towards showing that the coin is balanced in favor of H's, he might report on the subsequence corresponding to the trials 59 through 68 and ignore the other the data from the random sample of 200 coin tosses.

The sequence of H's and T's in Table 4.1 looks a little more regular than I expected. However, it is what the simulation gave me on the first run. The sequence has 88 runs, which is less than expected under the model of 200 independent tosses of a fair coin. Under this model, the number of runs has expectation 100.5 and standard deviation 7.05 so that the observed number of runs is 1.77 standard deviations below expectation.

At the end of Section 2.1, we discussed how persons generally do not understand the features expected from independent random selections. As a specific example of what I mean I will tell you of an experience with a class in 1989. I gave a class of twenty-five senior level students the task of using their minds to simulate 200 tosses of a fair coin. Two students put down sequences that had numbers of runs 89 and 90; there were 5 with number of runs between 102 and 108 inclusive; there were 6 with number of runs between 109 and 115 inclusive; there was 1 with number of runs between 116 and 121; and there were 11 with number of runs 122 and over with a maximum at 164, a whopping 9.01 standard deviations above model expectation. The students showed more understanding in terms of the number of H's in a sequence of 200 coin tosses; their misunderstanding was in terms of the distribution of the H's and T's within the sequence.

I would be remiss in not reporting a very bad experience that I had in which I carelessly mentioned an approach to modeling variability. The approach was inappropriate for the application, and the experience led me to appreciate the danger from thinking off the top of my head and in discussing ill-conceived ideas with others. In 1970, I received a call from a systems analyst working in the Michigan Department of Transportation concerning an analysis that she had been asked to do. An anonymous letter had been received following a bridge design civil service examination indicating that two individuals, identified as A and B, were copying and talking during the examination. It turned out that the multiple choice answers given by A and B were the same for 117 out of 120 questions, including 22 common incorrect answers. Apparently, the systems analyst had been asked to quantify the strength of this evidence against the pair. Not a simple task!

In our conversation, I began to think of the Binomial model as a benchmark for understanding the extreme nature of such coincidence of answers. I discussed a model wherein A and B select their answers independently according to a random mechanism. The model for the mechanism would allow for the estimation of parameters related to the probabilities of selecting a correct answer and the probabilities associated with the distractors or foils. The analyst listened to my ramblings and constructed a model based on which the statistical significance for the frequency of common answers by A and B was an absurdly small number.

Sometime later and to my surprise, I received a call from the Michigan Department of Civil Service asking me to give testimony in support of the approach taken by the systems analyst in developing her model-based significance. I was embarrassed, and, in good conscience, I declined to get involved. (In the many years since that time, I have had the opportunity to develop a greater understanding of the variability in common answers and common wrong answers among pairs of students. On several occasions, I have given advice to professors dealing with questions of the strength of evidence of certain degrees of commonality in answer patterns. On one occasion, I agreed to a request to testify at a hearing involving an accused student; this was unpleasant duty.)

In the early 1970's, I gathered data and attempted to adjust for such effects as common background and studying together in understanding variability in frequency of common wrong answers. I have reached a conclusion that any overly specified model will be deficient in this application. I like an approach of the Educational Testing Service on SAT's, an approach that may still be used. On a given day, the SAT may be given at say, 150, locations. Selecting one test-taker at random from each site creates a pool of 150 persons; within this pool, there is apparently very little chance for collaboration between any pair of individuals. The $C^{150}_2 = 11,175$ comparisons of answer patterns provide an empirical distribution for features and patterns of common answers. If a question arises at any one site concerning what proctors observe to be copying by an individual and/or collaboration by a pair of individuals, then an index of commonality is created for the specific pair based on the empirical distribution. Roughly, the index adjusts for how good the students are; if the index value under consideration deviates more than 3.7 standard deviations in a given direction from the mean of the empirical distribution, the SAT score(s) may be invalidated for the suspect(s).

Note that the approach involves a confirmation of a direct observation of copying or collaboration by comparison of an index value for the pair with a distribution which supposedly captures the variability of the index across pairs known not to have collaborated. The direct observation is critical; it would be inappropriate to create a distribution for your class after the test and accuse those pairs that were extreme. With no cheating, there will still necessarily be extremes due to natural variability. Angoff (1974) and Buss and Novick (1980) are examples of literature wherein issues related to detecting collaboration on multiple choice examinations are discussed.

Section 4.3   Hypergeometric Distributions

The Hypergeometric probability distributions have three parameters: N = # of elements in the population, n = sample size, and k = # of elements in the population of a particular type, which we will call Type A. The Hypergeometric distribution is the sampling distribution for the number X of A's in a simple random sample of size n selected from the population. We let p denote the ratio k/N, the probability of selecting a Type A element with a single random selection from the population. For a Hypergeometric random variable X,

$$(4.3) \qquad p(x) = \frac{C^k_x \; C^{N-k}_{n-x}}{C^N_n}, \; x = 0, 1, \ldots, n$$

and

$$(4.4) \qquad \mu_X = np, \qquad \sigma_X = \sqrt{\frac{N-n}{N-1}} \; \sqrt{np(1-p)} \; .$$

We denote this distribution by H(N, k, n).

When using simple random sampling and a small sample size relative to the population size, i.e., with n/N small, the Hypergeometric distribution is well-approximated by the Binomial with # of Trials equal to n and chance of success on a single trial equal to p = k/N. For this reason, the Binomial distributions arise as sampling distributions in simple random sampling from a dichotomous population. (If the selections were made with replacement, the Binomial is the exact sampling distribution.)

Example 4.9   Michigan Lotto 47

The current version of Lotto offered by the State of Michigan pays a large multi-million dollar jackpot for matching all six numbers in a combination chosen at random from the C^{47}_6 = 10,737,573 combinations of six numbers from the set {1, 2, 3, . . . ,47}. When the player specifies his/her six numbers, this specifies a subset of k = 6 elements. Subsequently, the Michigan Lottery chooses a simple random sample of size n = 6 from the population of size N = 47, the "winning numbers". Let X denote the number of elements in the randomly chosen combination that come from the player's combination. The player will win or share the jackpot if X = 6 , and will receive some payoff if X = 5 or X = 4. From the definition of the Hypergeometric distributions, we see that here X follows the distribution H(47,6,6):

Table 4.2. Michigan Lotto 47.

x	$C^6_x C^{41}_{6-x}$	p(x)
0	4,496,388	$4.1875 \ (10^{-1})$
1	4,496.338	$4.1875 \ (10^{-1})$
2	1,519,050	$1.4147 \ (10^{-1})$
3	213,200	$1.9856 \ (10^{-2})$
4	12,300	$1.1455 \ (10^{-3})$
5	246	$2.2910 \ (10^{-5})$
6	1	$9.3131 \ (10^{-8})$
Total	10,737,573	1.00000

Covering all combinations with $10,737,573 is not likely to be good bet. Of course, this depends upon the present value of the jackpot and the risk of sharing that jackpot with other players, which itself depends upon the total number of players and their choice patterns. (On Saturday, July 14, 1990, the jackpot had grown to over $21,000,000, but present value considerations and the risk of multiple winners made it a poor bet to cover all combinations. It turned out that there was one winner!)

In playing the Lotto, it would be useful to know the choice patterns of people playing the game. If enough people use birthdays in selecting their numbers, then the choices above 31 are better than those between 1 and 31, inclusive. I believe that the choice patterns of people playing Michigan Lotto are not available to the general public; this information should be either kept securely confidential by the State or made public information. ■

Example 4.10 Contested Elections

Consider an election involving two candidates A and B. Let a and b denote the numbers of votes cast for A and B, with a > b, and suppose that n illegal votes are included in the total N = a + b. Based on knowledge of a, b, n ≥ a − b, and with "no evidence" to indicate for whom the illegal votes were cast, should the election (of A) be declared void?

Finkelstein and Robbins (1973, p. 242) propose the following approach to resolving the question involving probabilities calculated from the Hypergeometric distribution:

> Consider all votes cast in a primary election as balls placed in an urn: black balls which predominate are those votes cast for the winner; white balls are for the loser. A certain number of balls representing the irregular votes are then withdrawn at random from the urn, an operation which corresponds to their invalidation. What is the probability that, after the withdrawal, the number of black balls no longer exceeds the number of white? Note the key assumption that the balls are withdrawn at random, i.e., that each ball has the same probability of being withdrawn. In terms of the real election situation, each voter is deemed to have the same probability of casting an invalid vote. This assumption will of course be untenable if evidence of fraud or patterns of irregular voting indicates that a disproportionate number of improper votes were cast for one candidate. But in the absence of such evidence, the assumption of random distribution of the improper votes is warranted. Whether or not mathematics is used to assess the probabilities, some implicit or explicit view as to the pattern of irregular voting seems inevitable. The assumption that each voter had an equal probability of casting an improper vote is the only neutral and non-arbitrary view that can be taken when there is no evidence to indicate that the probabilities are not equal. Thus in *Ippolito, DeMartini,* and other cases, where there was no evidence to disturb the assumption of randomness, the mathematical probability analysis depicted by the urn model is a correct expression of the intuitive probability used by the Court of Appeals in formulating the burden of proof standard for a new election.

Consider the case *Ippolito* in which a = 1,422, b = 1,405 and n = 101. In this election, A won the election based on the recorded vote by a margin of a − b = 17 votes. If n = 101 votes were removed by simple random sampling from the N = 2,827 votes, then the number X coming from A's vote would have to be 59 or greater in order to create a tie or to reverse the election. Under the Hypergeometric distribution H(2827,1422,101), μ_X = 50.804, σ_X = 4.935 and P(X ≥ 59) = .0592. If a judge hearing this case believes that this method is useful in assessing the likelihood of the election being reversed but for the illegal votes and that the calculated probability is substantial, he/she may void the election and call for a new election. Gilliland and Meier (1986) critique the Finkelstein and Robbins approach and suggest alternatives.   ■

Section 4.4   Poisson Distributions

The Poisson probability distributions have one parameter, λ. For a Poisson random variable X,

(4.5) $p(x) = \dfrac{e^{-\lambda}\, \lambda^{x}}{x!}$, x = 0, 1, 2, ...

and

(4.6) $\mu_X = \lambda, \quad \sigma_X = \sqrt{\lambda}$.

We denote this distribution by $P(\lambda)$.

The Binomial distribution B(n,p) is well-approximated by the Poisson distribution $P(\lambda)$ with λ = np if n is large and p is small. It follows that the number of chocolate chips in a chocolate chip cookie created from the random sampling of one ounce of dough from a mixing bowl that contains 300 chips and 50 ounces of mixed dough approximately follows the Poisson distribution P(6). This can be seen by regarding the mixing process as random placement of the 300 chips in the volume of dough so that the number of chips in a given volume amounting to one ounce, is like the result of n = 300 trials with chance of success p = 1/50 on a single trial. Therefore, the variability in number of chips in a cookie is described by a probability distribution with mean or expected value 6 and standard deviation $\sqrt{6}$ = 2.45. The probability for zero chips in a random cookie is e^{-6} = .0025 from P(6) and is $(49/50)^{300}$ =.0023 from B(300,1/50).

We discussed in Sections 2.1 and 4.2 that persons have a difficult time simulating independent tosses of a fair coin as evidenced by a production of too many runs of H's and T's in the sequence. In a sequence of H's and T's, a large number of runs is associated with the average length of the runs being small. Therefore, persons tend to believe that coin tosses tend to not exhibit long runs of H's and long runs of T's.

However, in fact, independent random selections tend to produce more clustering in time, as evidenced by long run's of H's and T's in coin tossing, and more clustering in space, as in the case of the points randomly distributed in a region in space. The following exercise involves a random distribution of points.

Exercise 4.4. Random Spatial Patterns. I used Microsoft QuickBASIC 4.0 to generate independent placement of 400 points at random in a square 320 × 320 region. Figure 4.3 shows the result for one quadrant

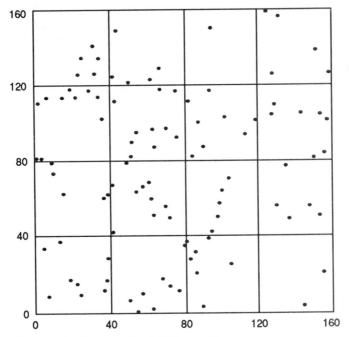

Figure 4.3 Random Spatial Distrubution

of the square which has been divided into 40 × 40 squares to help with the visualization.

Suppose that the count X in any one specified 40 × 40 square region is random and has the Binomial distribution B(400,1/64), which is approximately Poisson P(6.25). Calculate the average and the standard deviation of the 16 counts for the 16 subregions shown in Figure 4.3 and compare the results with the mean 6.25 and the standard deviation 2.50 for the Poisson probability distribution P(6.25). You may think of this as 400 chocolate chips being placed independently at random in a 320 × 320 sheet of cookie dough. This sheet is cut into 4 squares 160 × 160 for baking, and is then cut into pieces 40 × 40 for serving. The servings pictured in Figure 4.3 have from 1 to 10 chocolate chips in them. ■

Example 4.11. Colorant Content of Bottles

In teaching a short course at a company, I asked the participants to attempt to use the theory and methods developed in class to solve some of the problems that they were facing in their daily work. Most every problem that was chosen dealt with understanding the variability of outcomes in production processes.

One problem brought to class concerned the variability of colorant content in bottles blown by an injection molding device. The colorant provides the red color for the bottle; engineering specifications for a bottle stated that the colorant content should be between 2.0% and 2.4%. Colorant content below the lower limit produces an unacceptable pale red color while content above the upper limit is excessive in regard to expense and, when extreme, it produces a brittle plastic bottle.

Each bottle is made from a mixture of red (colorant) pellets, white (virgin) pellets that carry no colorant, and regrind material consisting of tags coming off plastic bottles that had been previously molded. The tags have been put through a grinder.

About 176 grams of the mixture of material make-up one bottle. The red pellets are uniform in shape; for our present purposes, we assume that each pellet weighs .05249 grams and has 47% colorant content. We also assume that regrind makes-up 35% of the material by mass and that regrind is 2.20% colorant. A calculation shows that a mixture that achieves 2.20% colorant will consist of 5.3549 grams of red pellets, 109.0451 grams of white pellets and 61.6000 grams of regrind. We calculate that the recipe calls for 102 red pellets per bottle.

The solid material is continuously mixed in a large vessel that feeds the molding device. The model of random placement of red pellets in the mixture suggests that the number of red pellets in 176 grams of the mixture follows the Poison distribution $P(102)$. The expected number is 102 and the standard deviation is 10.1.

This is rather excessive variation for this application. Consider a bottle receiving only 82 red pellets (4.3042 grams), a nominal level of 109.0451 grams of white pellets and an above nominal 62.6507 grams of regrind. This bottle will have only 3.4013 grams of colorant, which is 1.93% by mass and below the engineering lower specification level for colorant content. Moreover, if X is the number of red pellets in a random selection of 176 grams of the mixture, the chance of 82 or fewer red pellets is $P(X \leq 82) =$ .024 under the $P(102)$ distribution. Thus, the Poisson model suggests that 2.4% of the bottles produced will have colorant below the lower specification limit under fixed values of the virgin and regrind material. Therefore, in the factory environment we will not be surprised by the production of a high percentage of bottles that are outside of specification bottles.

If the red pellets were replaced by pellets one-fourth of their mass, then the number per bottle would follow a Poison distribution with $\lambda = 408$ under random placement in the mixture. Here the standard deviation would be 20.2 or 5.0% of the mean. This coefficient of variation is less than that resulting from use of the larger pellets, namely, 9.9%. Less variation in colorant content per bottle would result.

These mathematical analyses do not take into account many factors that would have to be considered before changing the pellet size. The engineering considerations would have to address mass segregation and other basic physical phenomena in the mixing vessel and the tubes that transport the material from the mixer to the blow molder.

However, I do believe that the above fairly simple mathematical analysis shows why the company was having so much difficulty in producing bottles to specification. We will revisit this example in Section 5.4 where we will incorporate all the sources of variability in discussing an error transmission formula and the capability of a process. ■

Section 4.5   Normal Distributions

The Normal probability distributions have two parameters, μ and σ. For a Normal random variable X, the probability density function is

$$(4.7) \qquad p(x) = \frac{\exp\left\{-\tfrac{1}{2}((x-\mu)/\sigma)^2\right\}}{\sqrt{2\pi}\,\sigma}\ ,\ -\infty < x < \infty$$

and

$$(4.8) \qquad \mu_X = \mu, \qquad \sigma_X = \sigma\ .$$

We denote this distribution by $N(\mu,\sigma)$.

The normal density is the familiar bell-shaped curve that is symmetric about the mean μ and has inflection points a distance σ from μ. (An inflection point is a point on a curve where the curve changes from being cupped upward to being cupped downward.) Figure 4.4 provides graphs of the densities for the distributions $N(0,1)$, $N(3,1.5)$ and $N(6,3)$.

The normal distributions play a central role in statistics because of the fact that the sums and averages of independent random variables will tend to have a distribution that is approximately normal. Precise statements of this phenomenon make-up the *Central Limit Theorems* found in advanced treatments of probability and statistics. Some of the Binomial, Hypergeometric and Poisson distributions are approximately normal with matching means and standard deviations.

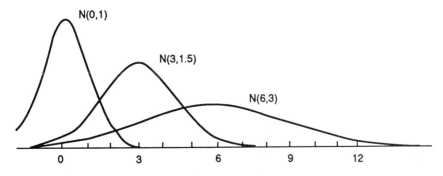

Figure 4.4 Three Normal Densities

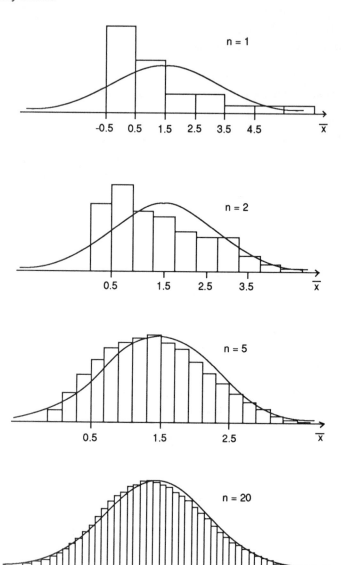

Figure 4.5 Distribution of $\overline{X}$ for Various n

I will illustrate the CLT phenomenon using software created by Erickson (1987). It plots the probability histogram for the average $\overline{X} = (X_1 + X_2 + \ldots + X_n)/n$ of independent random variables, all having the same discrete distribution on $\{0, 1, 2, \ldots, m\}$, $m \le 19$. It also superimposes a plot of the density function of the normal distribution that has the same mean and standard deviation as $\overline{X}$. Figure 4.5 shows the result for the discrete distribution

x_1	0	1	2	3	4	5	6
$p(x_1)$	.40	.25	.10	.10	.05	.05	.05

and various n. This discrete distribution has mean 1.5 and standard deviation 1.77482. The distribution of $\overline{X}$ has mean 1.5 and standard deviation $1.77482/\sqrt{n}$.

Note that as n gets larger the approximation by the normal gets better. By looking at the horizontal scales, you can see that as n gets larger, the probability distribution of $\overline{X}$ concentrates more closely to the mean.

When we model the distribution of a quality characteristic, we will often use the normal distribution. In a common cause system, the characteristic will deviate from its mean as the result of the actions of many independent sources of error.

Motorola (1988) describes its so-called Six Sigma Mandate as its goal for 1992 to have the distribution of quality characteristics centered on the target with σ so small that the range $\mu \pm 6\sigma$ falls within the engineering specifications. It goes so far as to print the probability $(2(10^{-9})$, i.e., 2 in a billion) of a normal variable falling outside the range $\mu \pm 6\sigma$ as an indication of the relative frequency of nonconforming product under this mandate. (It is hard to believe that any probability model is very useful in prediction in this sort of extreme use, but a model may be useful in regard to motivation and understanding.)

Section 4.6 Mixture Models and More Advanced Concepts

I will now discuss the application of the Binomial distribution in estimating the true vote for a certain contested election after giving some background information on contested elections in general. (I borrow freely from Gilliland (1984) in describing this experience.)

The result of a close election may be challenged on the basis of irregularities and the outcome may ultimately be decided by the courts. A given court may uphold the result, reverse the result, invalidate the entire election, or offer any of a number of remedies. In general, an election will not be overturned on the basis of a mere mathematical possibility that the results would be reversed in the absence of irregularities. Being reluctant to disenfranchise the valid electors unjustly, the courts have sometimes required that a challenger quantify the uncertainty that is created by the irregularity.

Example 4.12 The Flint Mayoral Election

I discuss the contested election held in the City of Flint, Michigan on November 4, 1975. Flint held the election for the office of Mayor, wherein a mayor was to be elected for the first time under a new city charter. The vote for mayor after the official canvass showed James Rutherford the winner over Floyd McCree with a margin of 206 votes (Table 4.3).

Table 4.3   Recorded Vote, 1975.

Candidate

Precinct	McCree	Rutherford	Totals
51	202	253	455
52	174	117	291
SubTotal	376	370	746
Other	20,099	20,311	40,410
Total	20,475	20,681	41,156

The breakdown for Precincts 51 and 52 is given because these vote totals were disputed due to a mix-up in the voting devices in those precincts. The vote totals from the other 143 precincts and absentee voters were not disputed.

Voting was done in Flint with punch cards. Here each voter at a precinct received a punch card which he or she took to a voting booth. In the booth was a voting device that worked as follows. The punch card was positioned to be punched by placing it in a slot. A booklet, which was fixed to the voting device, listed for each office the names of the candidates, with a hole beside each name. The individual voted for the candidate of choice by punching through the hole with the stylus that was provided. This removed a square (chad) from the punch card. All persons voting at a given precinct deposited their punch cards in the same receptacle as they were leaving the room in which the voting devices were located. The votes were subsequently counted by computer.

Precinct 51 in Flint had five voting devices and Precinct 52 had four voting devices. Rutherford was to be listed first among the two candidates for mayor in each of the booklets placed in the voting devices of Precinct 51. McCree was to be listed first in each of the booklets placed in the voting devices of Precinct 52.

By mistake, the election officials placed one booklet in Precinct 51 that should have been in Precinct 52, and one booklet in Precinct 52 that should have been in Precinct 51. The result was that each vote cast in the voting booth with the wrong booklet in either precinct was reversed, that is, recorded for the other candidate. The commingling of punch cards from all voting devices within a precinct made it impossible to distinguish which or how many votes were reversed within Precincts 51 and 52. The error was discovered after the polls had closed and the devices disassembled so that not even the locations within the polling places of the voting booths with reversed booklets were known.

A total of 746 votes were cast in Precincts 51 and 52 of which 376 were recorded for McCree. Because of the mix-up in booklets, it is possible that the total vote cast for McCree in these precincts was at least the 479 necessary to overcome the undisputed lead of 212 that Rutherford had across the other precincts. On this basis, Floyd McCree challenged in the courts the decision by the Board of Canvassers to declare James Rutherford the winner.

In Precinct 51, one of the five booklets had a reversed assembly. Suppose that each voter who votes in Precinct 51 selects the device with the reversed assembly with probability 1/5. Suppose that the voters make independent selections of the device on which they cast their votes and let θ_1 denotes the unknown number who voted for McCree. With this model, the recorded vote for McCree in Precinct 51 is $Z_1 = X_1 + Y_1$, where X_1 and Y_1 are independent random variables with distributions $B(\theta_1, 4/5)$ and $B(455 - \theta_1, 1/5)$, respectively. Moreover,

$$\hat{\theta}_1 = (Z_1 - 455/5)/(3/5)$$

has expected value θ_1 (is an unbiased estimator of θ_1) and has standard deviation 14.2.

In Precinct 52, one of the four booklets had a reversed assembly. Suppose that each voter who votes in Precinct 52 selects the device with the reversed assembly with probability 1/4. Suppose that the voters make independent selections of the device on which they cast their votes and let θ_2 denotes the unknown number who voted for McCree. With this model, the recorded vote for McCree in Precinct 52 is $Z_2 = X_2 + Y_2$, where X_2 and Y_2 are independent random variables with distributions $B(\theta_2, 3/4)$ and $B(291 - \theta_2, 1/4)$, respectively. Moreover,

$$\hat{\theta}_2 = (Z_2 - 291/4)/(2/4)$$

has expected value θ_2 and standard deviation 14.7.

It follows that $\hat{\theta} = \hat{\theta}_1 + \hat{\theta}_2$ has expected value θ, the total true vote for McCree from Precincts 51 and 52. Suppose that $\hat{\theta}_1$ and $\hat{\theta}_2$ are independent; then $\hat{\theta}$ has standard deviation 20.5.

The observed values of Z_1 and Z_2 were the recorded votes 202 and 174, respectively, from which $\hat{\theta}_1 = 185$ and $\hat{\theta}_2 = 202.5$, or an estimated total 387.5 for McCree for Precincts 51 and 52. (Subtraction from the total 746 cast for the office in these precincts yields the estimate 356.5 for Rutherford.) McCree would need 479 votes or more from Precincts 51 and 52 to overcome his deficit in the other precincts. Since the standard deviation of the estimate is 20.5, the model suggests that McCree was, in all likelihood, the loser of the election for mayor.

The assumptions of the model for the selections of voting devices by voters may be attacked as unrealistic. Subsequent to the hearing of the case, I was able to gather data on the voting distributions among electronic voting machines in a few precincts in Michigan to see whether the assumption was at all realistic. What data I could obtain, suggested that the Binomial model is questionable; apparently voters have a tendency to take the most convenient voting machine after receiving their ballot. This implies that during times when the precinct is not busy, the device closest to the position where the ballot is handed out will collect a more than average share of the votes. (On the other hand, during a busy period on election day, the votes will tend to distribute themselves uniformly among the devices with less variability than anticipated by the model that specifies independent selections of devices.) In the Flint case, the voter turnout was considered relatively light in the precincts with the reversed assemblies. Therefore, persons would tend to vote predominantly on one or two or three devices in each precinct. A mixture model approach that regards the probability of selection of the reversed assemblies as coming from distributions of possible values has merit for modeling the clustering by voting device and increased variability. I refer the reader to Gilliland and Meier (1986, p.409) for more on this point. ∎

Postscript to Example 4.12

The Board of Canvassers declared Rutherford the winner of the November 4, 1975 election on the basis of the 206 vote margin reported in the bottom line of Table 4.2. The ensuing court case was heard in Circuit Court of the County of Genesee on February 20, 1976. Plaintiffs presented very questionable models to show that McCree was the actual winner of the November election with a likelihood as great as 1 in 5. Plaintiffs sought relief in the form of a new city election. Defendants presented the analysis of Example 4.12 in a attempt to establish that Rutherford was the actual winner, in all likelihood.

On March 4, 1976 the Court ruled in favor of a special mini-election and in the Order indicated that "those voters are instructed to cast their ballot in the same way" (as cast in November). The letter to individual voters phrased it, "You are instructed to recast your ballot at this rerun election so that we may ascertain how you voted on November 4, 1975." McCree won 349 votes and Rutherford received 312 votes in this special election establishing Rutherford "firmly" in office with a citywide margin of 175 votes in a total of over 41,000 votes cast. ∎

Section 4.7 References

Angoff, William H., "The Development of Statistical Indices for Detecting Cheaters," *J. Am. Statist. Assoc.,* **69**, 44–49, 1974.

Bickel, Peter J., Hammel, Eugene A. and O'Connell, J. William, "Sex Bias in Graduate Admissions: Data from Berkeley." In Fairley, William B. and Mosteller, Frederick (Eds.), *Statistics and Public Policy,* Addison-Wesley, Reading, MA, 113–130, 1977.

Buss, William G. and Novick, Melvin R., "Indices of Cheating on Standardized Tests: Statistical and Legal Analysis," *J. Law and Education,* **9**, 1–64, 1980.

Erickson, Roy E., Statistical Software, *Michigan State University,* 1987.

Ethier, S. N., "Tests for Favorable Numbers on a Roulette Wheel," *J. Am. Statist. Assoc.,* **77**, 660–665, 1982.

Fabian, V. F., Statistical Software, *Michigan State University,* 1986.

Finkelstein, Michael O. and Robbins, Herbert E., "Mathematical Probability in Election Challenges," *Columbia Law Review,* **73**, 241–248, 1973.

Gilliland, Dennis C., "Probability in a Contested Election," *The UMAP Journal,* **V**, 1 10, 1984.

Gilliland, Dennis C. and Meier, Paul, "The Probability of Reversal in a Contested Election," In DeGroot, Morris H., Fienberg, Stephen E. and Kadane, Joseph B. (Eds.), *Statistics and the Law,* Wiley, New York, 391–411, 1986.

Meier, Paul, Sacks, Jerome and Zabell, Sandy L., "What Happened in Hazelwood (Statistics, Employment Discrimination, and the 80% Rule)," In DeGroot, Morris H., Fienberg, Stephen E. and Kadane, Joseph B. (Eds.), *Statistics and the Law,* Wiley, New York, 1–40, 1986.

Motorola, "Statistical Process Control," A Document of Motorola, Inc., Motorola Literature Distribution, Phoenix, AZ, 1988.

Rao, C. Radhakrishna, *Statistics and Truth: Putting Chance to Work,* Council of Scientific & Industrial Research, New Delhi, India, 1989.

Zeisel, Hans, "Dr. Spock and the Case of the Vanishing Women Jurors," *University of Chicago Law Review,* **37**, 1–18, 1969.

Statistics in the Quality Movement

In this chapter, I will discuss experiences and examples of the application of statistics in support of quality and productivity improvement. There is a wide-spread interest in statistics and statistical thinking today. Much of this is due to the impact and acclaim of Dr. Deming. Apparently, managers, engineers and workers are seeing a benefit from the use of statistical methods and thinking as never before.

In Section 5.1, I will discuss the idea of continuous improvement. Dr. Deming lists the creation of constancy of purpose to improve as the first of his Fourteen Points.

In Section 5.2, I will discuss Statistical Process Control (SPC) and one of the funnel experiments that Dr. Deming uses to illustrate the benefits derived from using SPC as opposed to the over-adjustment of processes in an attempt to control variation. I will use a chart showing the colorant content of molded plastic bottles to illustrate the principles learned from the funnel experiment.

In Section 5.3, I will discuss the concept of capability of a process and the indices that are popular measures of capability.

In Section 5.4, I will present the concept of error transmission. It is important to get a feel for the tools used to understand and model the variability that results when random phenomena combine from various sources in a multi-component assembly or design, and when they are transformed through various functions. The colorant experience will be revisited to show how an understanding of the Poisson distribution and an error transmission formula shed light on a lower bound on the variability in the colorant content of bottles.

Section 5.1 Continuous Improvement

In helping to organize a conference on the teaching and use of statistical theory and methods, I received calls from persons in industry, government and academia seeking to register and expressing a keen interest in the teachings of Dr. Deming, the keynote speaker. One call was particularly interesting and informative in regard to the zeal and fervor that has been unleashed here. A woman from the State of Michigan Department of Commerce told of an evolution of thinking in her department. First, the Department recognized the importance of quality in the marketplace and initiated programs to promote quality as a critical dimension of doing business today in Michigan. Second, they came to realize that the Department itself was not applying the principles underpinning the quality movement to its own operations. The Department chose the route of hiring consultants to teach them. As Department personnel listened to a consultant expound the virtue of control charting, one asked the consultant what characteristics he was charting in regard to his performance and the processes of which he was a part. To the dismay of the listeners, the consultant replied that he had not the time to do this, thus, leading to questions of his credibility and sincerity. The fact that the caller conveyed this message to me in a first conversation speaks to the deep interest and commitment that she has in improving quality of service in the Department of Commerce. It caused me to ask myself what I have done lately in monitoring and improving the processes for which I take some responsibility.

The eminent teacher, writer, research Professor C.R. Rao takes note of the importance and significance of statistical process control. Rao (1989, pp. 101–2) states that

> "In *industry,* extremely simple statistical techniques are used to improve and maintain the quality of manufactured goods at a desired level. . . . Indeed, there has rarely been a technological invention like *statistical quality control*, which is so wide in its application yet so simple in theory, which is so effective in its results yet so easy to adopt and which yields so high a return yet needs so small an investment."

I believe that most people think mainly about manufacturing and manufactured products when they think about quality, productivity and statistical process control. Yet the guru's in this area have stressed that the importance and usefulness of statistical thinking in regard to quality and management extend to the service industry, government and the schools. More and more managers and workers in these sectors of our society are becoming aware of the teachings of Deming, Taguchi, Ishikawa, Imai, Box, Crosby, Juran and others.

Each one of us can give examples of bad service but few of us seem to have any idea of how to effect lasting improvements in the way we serve our own customers, be they students, colleagues at work, the citizenry of the State. I wrote the following to be posted in my department after experiencing a particularly bad run of service over a short period of time in the Summer of 1989.

July 6, 1989

To whom it may concern:

I have experienced the following examples of lack of quality in service within the past few weeks. I guess that these experiences represent about 15% of the total service contacts that I have made over this period.

1. At XXXXX on Grand River Ave. in East Lansing a clerk could not figure the change coming from a \$10 bill with the charge being \$8.65. He asked his co-worker to help; the co-worker did quickly respond with the correct answer without the use of a computer or paper and pencil. (The clerk with the weakness in arithmetic claimed to have taken STT 315 from me in Fall 1988 and to have received the grade 3.0 He stated that he did worse in MTA 317.)

2. In ordering at XXXXX in Bowling Green, Ohio, I asked that the large burger be cut in two. It came back cut into two portions in the ratio of about 85/15 according to weight.

3. In ordering through the drive-up at XXXXX in Fowlerville, I asked for coffee (half decaf and half regular), two creams and a biscuit sausage sandwich with mustard. I drove away to discover that a packet of mustard was not provided. I will never know whether the coffee was what I had ordered.

4. In ordering through the drive-up at XXXXX on Grand River Ave. in Okemos, I ordered several items including XXXXX. The order came but did not include the XXXXX. (At least, I was not charged for them.) Straws were not given for the soft drinks.

5. In using the drive-up at XXXXX in Okemos, I made a deposit in one account and my wife wrote a check for cash on another account. We drove away with an envelope with the cash. However, the clerk had forgotten to include a slip showing the deposit. Presumably, it will be discovered and sent through the mail to me later at some expense to the bank.

These experiences began to bother me. We must improve quality in the service industry. We must recognize good service and reward it; we must speak out against bad service.

I think that Dr. Deming believes that most everyone wishes to do a good job and will do so if given the opportunity. The above experiences gave me reason to reflect on Dr. Deming. It became clear to me that the clerk who could not make change was not well served by the course STT 315 that I taught in Fall 1988. That class of 250 students was so diverse in its abilities; the class was directed at persons of average talent, and it did not transform some of its inputs in a way that added value. The clerk would have been better served by another course or experience. We must do a better job at restricting admission to and teaching STT 315; the process called STT 315 must be improved.

Decreasing the degree of variability in the input into a process will tend to decrease the variability of the output; belief in this principle is dramatically demonstrated by the movement by companies to reduce their number of suppliers.

Several of the problems that I experienced at the drive-up windows probably resulted from the poor communication that is possible with the speaker systems that typically exist at such locations. In at least one instance, I know that the order was taken by a clerk who figures the charge and collects the money at one window. With this system, the customer is then passed to another window where the order is picked-up having been filled by a different clerk from the one who took the order. Perhaps, some special orders and requests are lost in the communication link between the two clerks if, in fact, both clerks are not hearing the order directly from the customer. What a wonderful opportunity for a good manager to improve the system in which the workers are placed.

With these drive-ups, feedback from the customers to those giving the service is often missed. The manager can not hear the expletive deletes from the car as the customer opens the bag on I-96 two miles down the road.

I learned from these experiences that I should not make orders too complicated and should eat less fast food. I asked my colleagues for suggestions on how the service that I described can be improved. One response was to pay more money for these services jobs. I doubt that this is the solution. Rather, results will come when managers of these systems get turned-on to the Deming philosophy of continuous improvement.

How can this be done? It is said by Michael Leboeuf that the greatest management principle is "that you get what you reward". I tend to believe this and that adoption of the Deming philosophy will improve quality, so the question revolves around what reward system will promote the Deming philosophy of management. In summary, behavior can be shaped by a reward system, and, with the right reward system, service will be improved.

However, the rewards must be meaningful and have lasting value. I believe that simply giving workers more money or attracting a different group of persons through a marginal change in hourly rate is nearly useless.

It seems that to effect lasting change, the managers and their managers must adhere to Dr. Deming's first of fourteen points, namely,

> **"Create constancy of purpose toward improvement of product and service, with the aim to become competitive, stay in business and provide jobs."**

Working in an environment where there is this purpose will be its own reward and will help produce the kinds of monetary rewards that are necessary and appropriate. Dr. Deming has lots of deep insights into what actions can promote and maintain this constancy of purpose, can bring persons together as a team, and can produce job in the workplace. The employee appraisal system should recognize the value of teamwork and should not make the kinds of distinctions among employees that are counterproductive and divisive.

I am reminded of the story that Dr. Imai tells of the waiters and waitresses in a company in Japan. Here the company cafeteria was attended to by servers (waiters and waitresses) and part of their duty was to provide urns of tea at the tables. The servers noticed over a course of time that not all tea was consumed at certain tables and that the pattern was fairly stable. It seems that the non-tea drinkers and light-tea drinkers may have clustered at certain tables. After studying this, servers cut back on the amount of tea provided to these tables. Further observation showed that the customers (workers) were being satisfied under the new system. The servers wrote-up their findings, actions and results and showed management the report. They were recognized for their efforts in an appropriate way by the company. There was no large cash award, just a picture in the company newsletter or a plaque on the wall.

Dr. Imai says that the Japanese word for improvement is "kaizen", and that it can denote the small but steady improvement that results from the kinds of efforts that the waiters and waitresses made in the above example. Imai (1986) is a source of insights into a system that rewards and recognizes efforts at continuous improvement at all levels of a system.

I have been most fortunate to have had managers and fellow workers in most of my jobs that would recognize efforts at kaizen. However, I have never been part of a work culture that recognizes kaizen and promotes it to the degree that Dr. Imai reports for that company in Japan. I am afraid that too often workers who would attempt such "small improvements" as "saving a few tea leaves" would be the subject of laughter and derisive comments by their fellow workers and by management. What a shame!

I like the diagram that Dr. Imai gives that shows the increased opportunity for kaizen that is possible as a worker gains experience at a job. (See Figure 5.1.) By reading across horizontal lines from points on the experience axis, we see that the worker with more experience has a greater portion of his/her time on the job devoted to innovation and improvement and a lesser portion devoted to simply doing the job (maintaining the job). This makes good sense and suggests that, as we get more efficient in what we do, we should have more opportunity to improve what we do. If there is the feeling among the workers that there will be layoffs if management believes that fewer workers are needed because of increased efficiency, then the workers have little incentive to be more efficient. There will be less opportunity to practice kaizen.

Despite my own very positive impressions of my earlier work experience and the systems in which I was placed, I have heard of and seen instances where persons do their jobs in four hours and choose to use the other "working" hours for goofing-off. Worst of all, in some instances they are proud of being able to escape work!

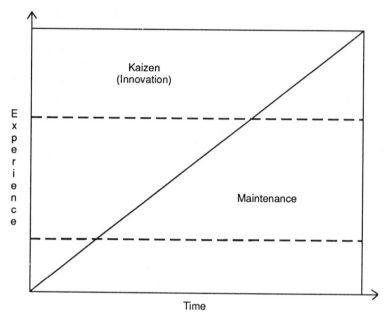

Figure 5.1 Diagram Showing Opportunity for Kaizen

Where is the joy in these workplaces? Managers, including those in top management, are responsible for the systems in which workers are placed and must do more to nurture a constancy of purpose in regard to innovation and improvement.

All of us can do more to encourage this constancy of purpose. Recently, I encountered an example of a most outstanding employee at a coffee shop on campus. I wrote a letter to his manager telling about the excellent service that the employee was giving. We should take the opportunities that we have to reinforce and recognize the excellence that is about us.

Section 5.2   Statistical Process Control

In Chapter 1, I told about the importance that Dr. Deming attaches to statistical process control (SPC). You can go to the library and find many books that explain a variety of control charting techniques. Introductory engineering statistics texts usually have a chapter devoted to the subject. The American Society for Quality Control (ASQC) is the professional organization whose members are most active in promoting and using SPC. Its publications *Quality Progress, Journal of Quality Technology* and *Technometrics* carry many articles on the subject. We will not provide any detailed account of the variety of charting techniques and their properties, but we will illustrate some basic ideas that underpin their usefulness.

To understand SPC, we must first understand the notion of a *stable system* of variation. Generally, stable refers to stable over time, but this can be space or some other dimension. The simplest example of a stable system is one involving independent repetitions of a fixed random experiment, for example, a system that generates independent tosses of a given coin, one toss after another over time. With a stable system such as this, variation over time in the H's and the T's will be solely the result of the system that is in place, and there will be no special causes of variation.

Dr. Deming's classic example of a stable system is the repeated random sampling of beads from a urn. His famous Bead Experiment is discussed briefly in Chapter 1 and is discussed in greater detail in Walton (1986, Chapter 4).

With a stable system, features of aggregates of future outcomes can be predicted to within certain limits of precision. For example, in the stable system consisting of independent tosses of a given coin, if the coin is fair, we can predict that the number of H's in the next series of 100 tosses will be between 35 and 65, inclusive. (With $p = .5$, the $B(100, p)$ probability for this is .9982.) If p were not specified, it could be estimated from the results of a long series of independent tosses that begin the series of trials. Suppose that the first 3000 tosses show 1599 H's. Then p is estimated to be $1599/3000 = .533$, and, conditional on the model $B(100, .533)$, the probability of producing between 35 and 65 H's in the next series of 100 tosses is .9931.

The number of H's per 100 could be charted with a so-called np-chart with the purpose of detecting a change in the process were it to occur. The Shewhart 3σ limits are located a distance 3σ from a center. For a normal distribution, the interval $\mu \pm 3\sigma$ captures 99.73% of the probability. The $B(100, .533)$ distribution is approximately normal with mean 53.3 and standard deviation 4.989 so the Shewhart limits are LCL = 38.3 and UCL = 68.3. The $B(100, .533)$ based probability for this interval is $P(39 \leq X \leq 68) = .9975$.

The Shewhart limits are used as follows. In a series of observations on the process, any i for which either $X_i < 39$ or $X_i > 68$ is termed a signal that the process is *out-of-control*. If the process has not changed, the rate of such signals will be .25%, and these signals may be called *false positives*. The expected number of trials i between such false positives is $1/.0025 = 400$.

However, suppose that the process rate p shifts from $p = .533$ to $p = .65$ so that the distribution $B(100, .65)$ governs the number of H's in lots of 100. With this distribution, the probability of an out-of-control signal is .2331 so that the expected number of lots of 100 that would pass before the change is signaled is $1/.2331 = 4.3$.

Example 5.1   Charting Brown M & M's

A colleague, Laura Ghosh, showed me a newspaper article that gave the nominal percentage breakdowns of plain M & M candies and peanut M & M candies into various colors. She told of conducting experiments in her classes consisting of having students opening bags of these candies and counting the various colors in order to illustrate the bag-to-bag variability in percentages about nominal values. Recently, I used this experiment in a class to illustrate the construction of a control chart called an np-chart based on target values. I will present the results based on charting brown candies in 1.69 ounce bags of plain M & M candies. For your information, I first give the newspaper report of the nominal percentages for all of the colors and both types of M & M's.

Color (%)

	Brown	Yellow	Red	Green	Orange	Tan
Plain M & M's	**30**	20	20	10	10	10
Peanut M & M's	30	20	20	20	10	0

By chance, a student who had interned at the Mars Company was present in the class. He told us a little about the filling process. A long tray of candies is created with brown, followed by yellow, followed by red, . . . having been placed there according to nominal percentages. (I assume this split is created through metering by weight and not an actual count.) The tray is emptied into a large mixer and mixing of a very large quantity of product takes place before the filling operation into individual bags. Because the plain candies are uniform in shape and mass distribution, it is reasonable to assume that there is a random distribution of the brown candies among those being dispensed from the mixer into the bags.

The eighteen students in my class were each presented with a 1.69 ounce bag of plain M & M's; all bags were purchased at one store the morning of the class. The following is a breakdown of the counts by students for each color reported in an order determined by the seating arrangement.

Student

	1	2	3	4	5	6	7	8	9	10	11	12	13	14	15	16	17	18	Total
Brown	20	20	22	21	16	24	12	19	16	17	18	13	14	22	23	11	11	17	316
Yellow	11	10	11	9	12	6	15	7	8	16	12	11	14	12	13	12	13	11	203
Red	7	5	11	14	9	12	12	16	12	11	15	7	16	11	12	16	14	8	208
Green	4	10	5	2	4	5	6	4	6	8	2	6	8	3	5	12	5	7	102
Orange	9	5	2	6	10	5	3	4	6	6	2	11	4	5	5	3	7	5	96
Tan	5	8	6	7	6	6	9	6	8	0	8	8	2	5	1	3	6	9	103
Total	56	58	57	59	57	58	57	56	56	58	57	56	58	58	59	57	56	57	1028

Consider the bags as random samples from the process that is dispensing candies that are nominally 30% brown candies. If the process were

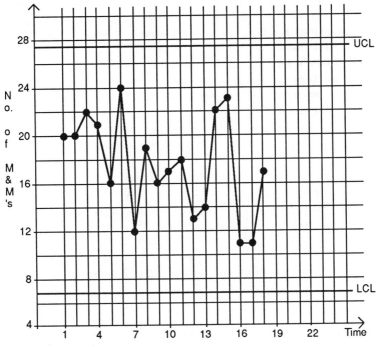

Figure 5.2 np—Chart for Brown M & M's

stable and performing as designed, the number X of brown candies in a fixed number n of candies would follow the Binomial distribution B(n,.30).

We see that the filling operation results in a variable total number of candies per bag. A more sophisticated model would model this additional component of variability. However, this would take us beyond what we wish to accomplish here. We use a fixed sample size n = 57 in the calculations yielding $\mu_X = 17.1$ and $\sigma_X = 3.46$, from which the 3 sigma control limits are LCL = 6.7 and UCL = 27.5. (The limits are based on the target value 30% and not the empirical percentage of brown candies in the aggregate from the eighteen bags, namely, p = 316/1028 = 30.7%.) Figure 5.2 is the control chart for the data.

Presumably, if this type of charting were done at the Mars Company, it would be done with the points in the order of the production. ■

As we have seen, a control chart is a plot over time of some characteristic of the output of the system. Its purpose is to present the results in an organized way, so that the persons responsible for maintaining the system in control can see the evolution of characteristic over time and can record its history. Based on the information in the chart at any time, the persons may take action (make adjustments to the process) or leave it alone. Certain prescribed statistical tests should be in place, such as Shewhart's, so that the decision to adjust the process or not is based on operationally defined standards and not the whims of the persons in charge.

I have emphasized on several occasions that people, using their own judgment and unaided by theory, are not good judges of randomness or stability; they must be helped with certain standard statistically motivated tests that are designed to detect changes in the process as quickly as possible. In all of this, the choice of procedures is based on the process knowledge that engineers and operators have as to the types of changes that are likely to occur, if any. Some procedures, such as the CUSUM procedures, are quicker than the Shewhart at signaling slow drifts in the mean of a process.

I am intrigued by the fact that often up to six tests for "out-of-control" are applied to a single control chart. I describe the tests below for an $\overline{X}$ – chart with the Zones as indicated in Figure 5.3.

Test 1. Any point beyond the control limits.

Test 2. 7 points in a row either above or below the centerline.

Test 3. A run of 7 points in a row moving up or moving down.

Test 4. 2 out of 3 points in a row in Zone 3 or beyond on one side of the centerline.

Test 5. 4 out of 5 points in a row in Zone 2 or beyond on one side of the centerline.

Test 6. 3 points in a row in Zone 2 or beyond on one side of the centerline.

	UCL
Zone 3	
Zone 2	
Zone 1	
Zone 1	
Zone 2	
Zone 3	
	LCL

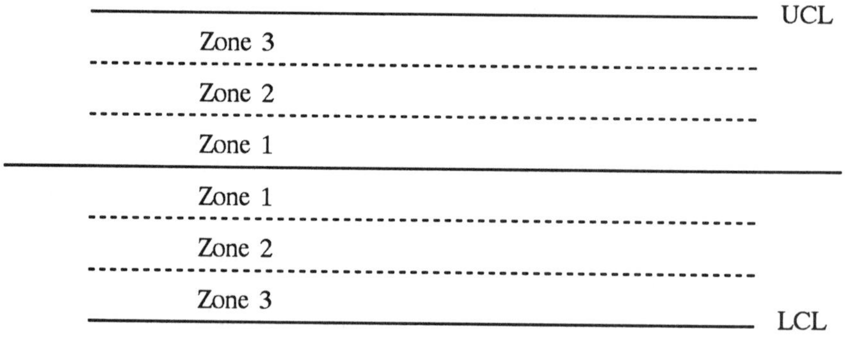

Figure 5.3 The Zones for an $\overline{X}$ – Chart

Consider an $\overline{X}$-chart with a subgroup size n and the model wherein the X_i's are independent random variables with the distribution $N(\mu, \sigma/\sqrt{n})$. We will assume for our purposes that the control limits are the exact $\mu \pm 3\sigma/\sqrt{n}$ so that the probability of a false positive at any time i, is .0027. Thus, the average run length (ARL) between these false positive coming from Test 1 applied to this perfectly stable process is 1/.0027 = 370.

Consider Test 2 applied to the same process. From Feller (1960, XIII.8.3) specialized to p = q = .5 and r = 7, the ARL between false positive coming from Test 2 is 127. When applying the battery of all six tests, I estimate that the ARL between false positive is less than 100. In the literature, you will find reports of studies of the rates of false positives and the powers of these tests applied individually and in batteries.

I have had a computer program written that simulates the evolution of an $\overline{X}$ chart with the $\overline{X}_i$'s independent random variables with the $N(\mu, \sigma/\sqrt{n})$ distribution. Control limits are set at null values $\mu_0 \pm 3\sigma_0/\sqrt{n}$. Students can set μ and σ values for the process, generate a history of several thousand $\overline{X}_i$'s, and see the result of the application of the six tests on the history. The choice $\mu = \mu_0$ and $\sigma = \sigma_0$ serves to demonstrate false positives. Other choices serve to demonstrate "true" positives, by which I mean, signals of "out-of-control" when the process parameters are different from those on which the process control limits were set. In the language of hypothesis testing, the probability of a false positive corresponds to α = Prob(Type I error), and the probability of a true positive corresponds to $1 - \beta = 1 -$ Prob(Type II error).

By working with this computer program, students quickly learn that, by increasing the number of tests, there is an increase in the rate of false positives when the process is in-control with the benefit being an increased rate of true positives when the process is out-of-control.

Ideally, in practice, operators, engineers and statisticians work together to tailor the package of tests for individual processes depending on the opinions as to the most likely types of special causes of variation and out-of-control situations and their likelihoods. In practice, applying all six tests will create so many out-of-control signals that process managers and operators are apt to discount them and to choose to investigate only a relatively few of the signals.

As quality is designed into more and more processes, one can expect the application of less stringent tests to the charts and that there will be less need for charting as well as for mass inspection and acceptance sampling.

You can readily see the advantages to producing and consuming product that is produced by a stable system. If I am a producer, it means that I need commit fewer resources to final inspection. Suppose that defective parts are being produced with independent trials at a rate of p = .001. Then as a consumer buying in lots of size 1,000, I would know that the number of detectives per lot follows a B(1000,.001) distribution. The expected number per lot is 1 with a standard deviation of approximately 1. (Of course, I could use sampling on the lot to produce enumerative information specific to the lot.) I could plan the system that used these parts having this information and would spend less money on incoming inspection of these parts.

More and more, companies are requiring that their suppliers furnish a history or pedigree of the process that produced the parts that they are buying. Often these take the form of control charts. There is increased communication between companies and their suppliers as part of vendor certification efforts by companies. The need for the operational definitions of terms and the standardization of statistical methods is apparent.

When a company audits its suppliers and potential suppliers, there is generally much attention to quality issues and the statistical records that speak to the quality and statistical control of the production processes of the vendors. General Motors has its Targets for Excellence Program; Ford and Chrysler have their programs. Each company may choose to weight characteristics differently, but each regards statistical process control and the fundamental commitment to continuous improvement as very important in its evaluation of suppliers.

Michael Martin, Director of the Industrial Development Institute located at Michigan State University, read of an experience that illustrates the amount of emphasis that there is in these audits on the commitment to continuous improvement. A major automobile company dropped a supplier that had possibly never sent it a nonconforming part in its past years of business with the automobile company. The supplier was using 100% final inspection to ensure that only conforming product was shipped. This is not good enough; rather it is symptomatic of a supplier that does not have quality and statistical control built into its processes. The supplier was not driving its processes to reduce variability; rather it was relying on an inefficient method to assure quality. It had no plan to improve what it was doing; no cost reductions were possible; no commitment to continuous improvement was observed. Protestations by the supplier that it had never shipped nonconforming product were ineffective; the supplier was dropped.

Of course, companies are anxious to reduce variability and to stay in business. Removing variation is not necessarily an easy task, and the uninformed are likely to be unselective in the variations they chase. This is why an understanding of variability and the principles that Dr. Deming enunciates is so critical.

Dr. Deming, in notes for his NYU Seminar in March, 1988, warns of wasteful attempts by some to search out special causes of variation when the data do not suggest such action be taken. He refers to action taken on a stable system in response to variation *within* the control (Shewhart) limits, in an effort to compensate for this variation, as *tampering*. He notes that such tampering will increase variation, not reduce it. He uses a "Funnel Experiment" to illustrate this phenomenon.

Example 5.2   Target Practice

I believe that you can visualize the phenomenon called "tampering" by thinking of an expert with the rifle who shoots at a target. Figure 5.4(a) shows the pattern of bullet holes resulting from n = 5 shots at the bullseye after the gun has been zeroed in to the extent possible to remove systematic error and has been placed in a vice. The variation that you see about the

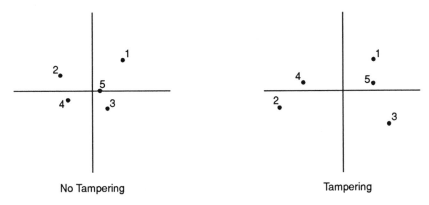

No Tampering   Tampering

Figure 5.4 Target Practice—(a) No Tampering (b) Tampering.

bullseye is due to variation in the powder loads from shot to shot, variation in the geometry and mass distributions in the bullets from shot to shot, variation in wind velocities from shot to shot, etc. These are sources of variation that concern the expert; he/she may load the shells and purchase precision equipment (powder, bullets, rifle, . . .) in order to reduce variation. This person is continually making fundamental changes and improvements in the system in order to reduce variation in the output so as to improve the results. This person realizes that he/she can not control the wind other than to choose days with as little wind if that is possible.

Dr. Genichi Taguchi has emphasized designed experiments with factors including controllable variables and the uncontrollable variables. The intent is to of find a combination of controllable variables (shells, powder, powder charge, bullets, rifle) that produce robust behavior in the presence of variation that can not be controlled (wind). His insights have had a tremendous influence on the practice of designed experiments. (See Taguchi and Wu (1979) and Taguchi (1986).)

In Figure 5.4(a), we see the result of what might be modeled as random variation about the aim point, which is the center of the target. On the other hand, in Figure 5.4(b) we see what would have happened with the same shots at the same times, if the person were to have adjusted the sights after each shot to compensate for the deviation of the preceding shot from the bullseye. Here you can dramatically see the effect of tampering in an attempt to compensate for variation for a system that is stable. Variation about the target bulleyes has been increased! ■

Example 5.3   Computer Simulation on Tampering

I will now illustrate the effect of tampering by providing the output of a simple computer simulation of a stable process. Figure 5.5 shows two series. In the upper series, you see the plot of outcomes for independent, uniformly distributed variables on the interval (−10,10), plotted across the

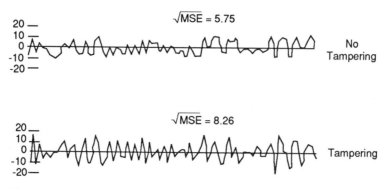

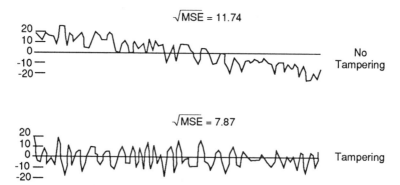

Figure 5.5 Tampering with a Stable System

Figure 5.6 Tampering with a Drifting System

points i = 0, 1, . . . , 80. The target value is zero. In the lower series, you see the result of tampering at i = 1, 2, . . . , 80 through the subtraction of the previous deviation from the aim point in an attempt to compensate for the variation. Since the variation is the result of independent and identically distributed random variables with mean 0, this attempt is fruitless and adds to the variation. An analysis shows that this tampering increases the standard deviation of the errors by a factor $\sqrt{2}$.

Above each series, I give the square root of the average of the squared deviations of the observations 1 through 80 from the target value 0. The square of this measure is called mean square error (MSE) in statistical jargon. Dr. Taguchi favors the evaluation of a distribution of quality characteristics through a MSE; when scaled to give a value of the loss, these are known as "Taguchi loss functions".

As we have seen, the tampering that Dr. Deming speaks of does not pay when the system is stable. However, if there is a drift in the process mean, this tampering may be helpful. Figure 5.6 shows the results of the no-tampering and

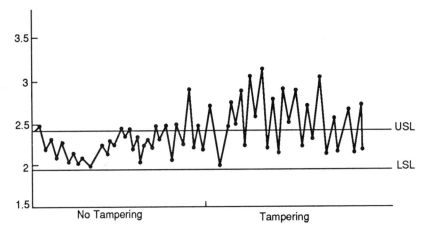

Figure 5.7 Effect of Tampering on Colorant Content

tampering on a series of independent uniform (–10,10) random variables with a drift added that shifts the mean linearly from +20.5 at i = 0 to –20.0 at i = 80. Above each series, I give the square root of the sum of squared deviations of the observations 1 through 80 from the target value 0. Here "tampering" has helped to reduce the deviations from the target. ■

We would have to get into concepts of engineering control systems in order to develop the principles that would be useful guides in the understanding of the advantages and disadvantages of various control systems. This is much beyond the scope of this book.

Example 5.4   Variability in Colorant Content of Bottles

In Example 4.11, we introduced an experience that concerns the excessive variability in the colorant content of certain plastic bottles. Dry materials are mixed to feed an injection molding device. The colorant pellets carry most of the colorant into the mixture to provide for the desired red color. Figure 5.7 is a rendition of the actual chart for the colorant percentages of single bottles taken off each shift of the production over a period of time.

At a time indicated on the horizontal time scale, the operators began a control strategy in which adjustments were made to compensate for estimates of the mean colorant content for the boxes of red pellets brought into production. The estimates had sampling error; the adjustments based on the estimates had the same effect that we see from tampering. ■

The proper approach has managers, who are responsible for a production system or process, supporting operators and engineers in their efforts at problem solving to remove the special causes of variation that are indicated by out-of-control signals from control charts. Stability can be achieved through such care and thoughtful maintenance of the system.

Is one to be content once stability is reached and maintained? No! Dr. Deming's first point implies that there must be continuous effort at the improvement of processes. The next example shows how there can be a direct monetary benefit to a company through an improvement in the form of reduced variability in a filling operation.

Example 5.5 Filling a 16 Ounce Package

Consider a filling operation with a 16 ounce net weight label claim. Suppose that the packages are shipped in lots of size 1,000, and that the product is covered by the NBS Handbook 133 (1984) Category B rules for label claim specifying that in a random sample of n = 30 from the lot, at most one package with less than 15.80 ounces is allowed. Suppose that the filler dispenses product into packages in an amount that follows a Normal distribution N(16.20,.20) and the amounts dispensed are independent, package to package. Then the chance that a single package is filled below 15.80 ounces is p = .02275, and, from the B(30,.02275) model, the chance that the sample contains more than one package filled below 15.80 ounces is .1485.

Improving the process by reducing the standard deviation to .05 ounces and saving product (for the company) by reducing the mean to 16.10 ounces, changes the chance that a single package is filled below 15.80 ounces to less than 1 in a billion. ∎

Example 5.6

I recently learned of a company that ran a filling operation for a product based on a target value $T = L + 1.28\sigma$, where L is the label weight and σ is an estimate of the process standard deviation of fill weight per bottle. This target value is set at 1.28 standard deviations above the net weight printed on the label because the government allows a rate of 10% for underfill and because the Normal distribution $N(T,\sigma)$ puts 10% of its probability below L.

This is cutting things too close and does not recognize the fact that errors in the direction of underfill may be more serious than errors in the direction of overfill. This is a model-based setup for the process and models should always be used with a degree of skepticism.

However, with the model, we see that deviations of the process mean μ about T will change the instantaneous rate of underfill. Small and stable deviations across time will result in underfill for about 10% of the bottles in aggregate. However, this is a very high speed line, and this company actually experienced a downward drift of process mean towards L. In a relatively short period of time it produced 20,000 bottles of which about 40% were filled to less than the indicated label weight. This was determined later by individual inspection at great expense to the company. ∎

Dr. Deming points out that the improvement of a stable system usually requires that fundamental change be made in the system. This is the responsibility of the managers. Such changes can be simple but fundamental; an example given by Dr. Deming involves increasing the level of lighting in a factory. A stable system of errors was changed to a stable system of errors with a lower average and less variation. More complicated fundamental changes may involve the introduction of new equipment and system design. I believe that, for the colorant example that we have been discussing, the least variability that can be achieved with the system when brought into control will still be much too large and wasteful. Perhaps, the fundamental change that is needed will involve the metering of materials into a mixing device in closer proximity to the blow molder. This is an engineering design change for the process; experiments in a laboratory would be useful in coming to a conclusion as to the fundamental changes needed to improve the process in a fundamental way.

Motivation and training are critical components of Dr. Deming's philosophy. William Conway, former CEO of the Nashua Corporation, speaks eloquently to the need for training of employees to provide them with basic statistical and problem solving skills, the tools necessary in the support of continuous improvement. He tells of a manager who brings two persons into his office to be instructed as to the rules of a contest between the two. The prize for the winner is an all-expense paid trip to Hawaii with the family, an extra week of vacation and a $5,000 bonus. Both employees are extremely well-motivated to win the contest. Then the contest is announced. The winner will be the first to screw a $1''$ screw into an oak plank. One employee is provided with a tool, a screw driver; and the other is not provided with any tools and is left with his thumbnails to do the job. We know who will win the contest. Tools and motivation are both needed!

I know of companies that reward district managers for efforts that are counterproductive to the company as a whole. For example, one packaged food company gives rewards to the managers with the greatest degree of customer satisfaction in their districts. (Here the term "customer" is used to mean the retailers of the food products.) The managers are responsible for placing orders with the company and with supplying their retailers with a packaged food that has a shelf-life of several months. However, district managers are supposed to share inventory of the product when they have a certain margin of inventory and another district manager is running short. However, the reward system encourages a district manager to be slow in processing requests for inventory from his/her fellow district managers in order to have the greatest chance for high customer satisfaction within his/her district. The global effect is to produce hoarding and inefficiencies for the company.

Another example has a company rewarding its plant managers each year whose plants have the fewest accidents. The plants are air processing plants with very few employees. For a given plant operating under stable conditions, the accidents may follow a Poisson distribution with a small λ. It would take many years of experience with a plant to estimate its λ with any degree of relative precision, that is, with a small coefficient of variation. Each year, some plant managers are rewarded for something that may best be attributed to random variation and not their actions. Because of the varying degrees of repair and age of the plants, this is a particularly divisive way to single out some plant managers over others and may even contribute to the under reporting of accidents. The question of operationally defining an "accident" may not even have been addressed.

Dr. Deming uses the notions of SPC to formulate fundamental ideas that are extremely important in the management of all types of systems and processes. The fundamental idea is so simple and yet so elegant. A manager and his team are responsible for achieving stability and maintaining stability in their systems. The manager is responsible for creating an atmosphere that will breed the philosophy of continuous improvement and for initiating and supporting efforts at fundamental changes to improve the stable processes. A fundamental change for a process that is out-of-control is less likely to have positive results.

As I was driving somewhere on February 16, 1990, I was listening to the popular Focus Program on WJR AM in Detroit. J. P. McCarthy was interviewing the Warden of the Jackson State Prison, the largest walled prison in America with some 5,000 inmates. I was impressed with the wisdom that the Warden demonstrated in responding to questions about drug testing of the prison inmate population. The Warden stated that, on a regular basis, random samples of about 100 prisoners were selected. He stated that the rates for positive test results averaged around 7%–8%, but, if something as extreme as, say, 20% were observed, a special investigation would be launched. He spoke with a confidence that led me to believe that he managed his system (the prison) with principles that Dr. Deming espouses. It was clear that he understood that it was his job to support efforts at maintaining control of the drug usage by ferreting out special breaches of security that would have led to excessive rates and out-of-control signals when they occur. It was his job to ensure that effort was not wasted at chasing variation that could easily be due to sampling error. It was his responsibility to make the fundamental changes in his system that would reduce the rate of drug usage in his prison from 7%–8% to 5%–6%, and then lower. Rather than chasing spurious variation that was due to sampling error and the common causes in the prison system, he was working on fundamental changes through improvements in prison's security, education and rehabilitation subsystems.

Exercise 5.1. A molding operation is used to mold a small plastic drinking cup. The weights (in grams) of the cups from a single cavity in the mold are charted over time. Every four hours, four consecutively produced cups are weighed. The results for 80 hours yield the following data from the $k = 20$ subgroups of size $n = 4$. Summary statistics for each group are also given.

Group(i)	X_{i1}	X_{i2}	X_{i3}	X_{i4}	$\bar{X}_i$	R_i
1	4.0668	3.9673	4.0872	4.0016	4.0307	.1199
2	3.8829	3.9087	4.1111	4.0622	3.9912	.2282
3	4.1426	4.0761	4.0821	3.9049	4.0514	.2377
4	4.1363	3.9739	3.9707	3.7567	3.9594	.3796
5	4.0394	3.7973	3.8831	4.0081	3.9320	.2421
6	4.0437	4.1064	4.0451	3.7522	3.9869	.3542
7	3.8278	4.0444	3.8845	4.0270	3.9459	.2166
8	3.9761	4.1700	4.0027	4.0151	4.0410	.1939
9	3.9137	4.0319	3.9013	4.0686	3.9788	.1673
10	3.9538	3.9298	4.1092	4.1003	4.0233	.1794
11	4.0525	4.2024	4.0120	4.1273	4.0985	.1904
12	3.9401	3.9118	4.0094	4.0112	3.9681	.0994
13	4.0232	3.8771	3.9390	3.9995	3.9573	.1261
14	4.0811	3.9111	3.9096	4.1197	4.0054	.2101
15	4.1690	4.1855	3.9043	3.9690	4.0570	.2812
16	4.0250	3.9967	3.8621	3.9260	3.9525	.1629
17	3.9840	4.2859	4.0208	4.0948	4.0964	.3019
18	4.0464	3.9690	3.8529	3.9354	3.9509	.1935
19	3.8645	3.9339	4.0085	4.0783	3.9838	.2138
20	4.1690	3.8833	3.9784	4.0371	4.0170	.2857

Estimate the standard deviation of X using the formula $\hat{\sigma} = \bar{R}/d_2$ with $d_2 = 2.059$. Plot the centerline $\bar{\bar{X}}$, the control limits $\bar{\bar{X}} \pm 3\hat{\sigma}/\sqrt{n}$ and the control chart of the 20 group averages. Are there any out-of-control points? ∎

Section 5.3 Capability Analysis

The future aggregate performance of a process that is stable can be predicted under the model that has been fit to the process. Then as long as the process stays stable and the model applies, the predictions will be usefully precise. Outcomes on individual parts or products can not be predicted, but aggregate performance can. In Section 4.5 we saw where Motorola used the Normal distribution for a variable quality characteristic to predict a very small rate of production of nonconforming product.

Two indices are in common use that describe how capable a stable process is relative to specific specifications for a variable characteristic. The engineering tolerance or simply "tolerance" is an interval running from a lower specification limit (LSL) to an upper specification limit (USL). If the data from the stable process indicate that the process mean is at μ and the process standard deviation for the variability of the characteristic of individual parts is σ, then the C_p index for the process is

(5.1)
$$C_p = \frac{USL - LSL}{6\sigma} .$$

The numerator is referred to as the *engineering tolerance* and the denominator is referred to as the *natural process tolerance*. A value for the C_p ratio of 1 or greater implies that the process distribution, if centered in the engineering tolerance region, will fit nicely in that region. If $C_p = 1$, the model using independent selections from the Normal distribution $N(C,\sigma)$ predicts that 99.73% of the product will be conforming to engineering tolerance. If $C_p = 2$, the prediction is 99.99%.

The C_p index indicates how capable the stable process is if it can be kept centered in the tolerance interval. Any drift of the mean μ away from the center $C = \frac{1}{2}(LSL + USL)$, will decrease the probability content of the tolerance interval, provided the standard deviation σ stays the same and the model $N(\mu,\sigma)$ applies.

An index which discounts the C_p index by taking into account that the process mean μ may not equal C is,

(5.2) $C_{pk} = \dfrac{2 \min \{\mu - LSL, \ USL - \mu\}}{6\sigma}$, if $LSL \leq \mu \leq USL$

and is undefined if μ is not in the tolerance interval. If $\mu = C$, then $C_{pk} = C_p$.

Exercise 5.2. Calculate the C_p and C_{pk} indices from the data provided in Exercise 5.1 assuming that LSL = 3.80 and USL = 4.30. Do you regard the process as very capable? Predict the rate of nonconforming product using the Normal model. ▪

Whereas the above capability indices measure how well a normal distribution fits inside an engineering tolerance, when applied to non-normal distributions, they imply little about the probability of producing nonconforming product. In an interesting series of articles, Gunter (1989a, 1989b, 1989c, 1989d) explores this point and other important points concerning the use of C_p and C_{pk} indices. These articles are a good source of information and insight into the effect of estimation errors in estimating these indices from process data.

I like the language in Motorola (1988, p.6) and the Ford Motor Company illustration found in its "Continuing Process Control and Process Capability Improvement." The illustration is reproduced below as Figure 5.8 with the kind permission of Ford.

> One of the goals of SPC is to ensure that a process is "capable". Process capability is a measure of its ability to consistently produce products that meet their specification requirements. The purpose of a process capability study is to separate inherent "random variability" from "special causes"

Process Stable

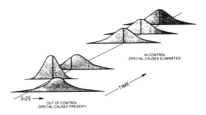

Process Control

Process "under control" - all special causes are
removed and future distributions are predictable.

Process Unstable

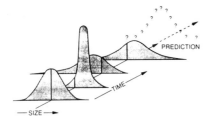

Process Capability

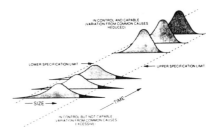

Source of illustrations – Continuing Process Control and Process Capability Improvement, Ford Motor Company

Figure 5.8 Impact of Special Causes on Process Predictability/Difference
Between Process Control and Process Capability

and then eliminate the most significant "special causes". Random variability
is generally present in the system and does not fluctuate. It may sometimes
be defined by basic limitations associated with the machinery, materials,
people's ability or manufacturing methods. Special causes come and go in
the system causing time-related variations in yield, performance or reliability.
Traditionally, special causes have often appeared to be random because they
were never examined closely enough. Figure 3, below, shows the impact on
predictability that special causes can have. Figure 4 shows the difference between process control and process capability.

A process capability study involves taking periodic samples from the
process in a controlled manner and charting the performance characteristics against time. Special causes are identified and engineered out.
Often, careful documentation of the process is key to accurate diagnosis
and successful removal of the special causes. In other cases, the causes
will remain unclear and experimentation will be needed.

Section 5.4 Error Transmission

It is useful to understand what results when various sources of variation combine or a random characteristic is transformed. For example, sup-

pose that Part A and Part B are assembled or packed end-to-end in Space C. Suppose that Part A is produced by a process that is in-control with respect to its length X_A and that X_A follows a distribution that has mean μ_A and standard deviation σ_A. With similar notations and assumptions in regard to B and C, we see that the difference

(5.3) $$X_D = X_C - (X_A + X_B)$$

is the excess space or room in Space C after Parts A and B are placed in the space. Thus, X_D is a critical characteristic in regard to the assembly and fit; if $X_D < 0$, then the parts would not fit into the space. Understanding the relationship of the distribution of X_D to those of X_A, X_B and X_C is important.

Suppose that X_i, i = 1, 2, . . ., n are independent (more generally, uncorrelated) random variables with means μ_i and standard deviations σ_i and that a_i, i = 1, 2, . . . , n are constants. Then, if $Y = \Sigma a_i X_i$,

$$\mu_Y = \Sigma a_i \mu_i$$

(5.4)

$$\sigma_Y = \Sigma \sqrt{a_i^2 \sigma_i^2}$$

These relationships are important in determining the effect of adding independent sources of variation that occur in multiple component assemblies. The familiar formulas for the mean and the standard deviation of an average of independent and identically distributed random variables are special cases.

Returning to (5.3), we obtain

$$\mu_D = \mu_C - (\mu_A + \mu_B)$$

and

$$\sigma_D = \sqrt{\sigma_C^2 + \sigma_A^2 + \sigma_B^2}.$$

If $\mu_A = 15.70$mm, $\sigma_A = 0.01$mm; $\mu_B = 14.30$mm, $\sigma_B = 0.02$mm; and $\mu_C = 30.20$mm, $\sigma_C = 0.05$mm; and X_A, X_B and X_C are independent, then

$$\mu_D = 0.20\text{mm}$$

and

$$\sigma_D = 0.055\text{mm}.$$

Suppose that the engineering tolerance for X_D has LSL = 0.10mm and USL = 0.50mm. Then the process for X_D has $C_p = 1.2$ and $C_{pk} = .6$. If X_D were normally distributed, the probability of a total failure $X_D < 0$ is .00014.

There are interesting computer software programs that engineers use in designing complicated assemblies involving many parts and various geometric configurations, for example, Variation Simulation Modeling by Applied Computer Solutions, St. Clair Shores, MI. The mathematics and statistics that underpin these programs involve the theory of projections of error distributions on planes and lines and various formulas appropriate for combining errors. The above example involving Parts A and B assembled into Space C is an extremely simple illustration of error analysis in one dimension. Computer support is essential for the analyses in two and three dimensions.

We now discuss what results from the transformation of a random variable X. A process in its most general form acts to transform inputs into outputs. The inputs can generally be classified into such categories as human, material, environment, . . . The most basic of models that captures the essence of a process has input X, process f and output $Y = f(X)$, where f denotes the function that maps or transforms the input X into the output Y. The input X and the output Y may each consist of many variables as components, that is, may be multidimensional.

In terms of notations with which we have become familiar, if X is numerical and the transformation is linear with $Y = a X + b$, then

(5.5)
$$\mu_Y = a \mu_X + b$$
$$\sigma_Y = | a | \sigma_X.$$

The function, $y = f(x) = a x + b$ is a linear function of x. Thus, (5.5) relates properties of the distribution of the linear transform $Y = f(X)$ to that of the variable X.

A smooth function $y = f(x)$ is approximated by the linear function. Specifically,

(5.6)
$$y \doteq f(x_0) + f'(x_0) (x - x_0)$$

in a neighborhood of the point x_0. In (5.6), f' denotes the derivative of the function f so that $f'(x_0)$ denotes the derivative or slope of the function f at the point x_0.

It follows that if X is random variable that concentrates its probability in a neighborhood of its mean and Y is the random variable $Y = f(X)$, then

(5.7)
$$Y \doteq f(\mu_X) + f'(\mu_X) (X - \mu_X).$$

From (5.5) and (5.7), we get the approximation formulas for the mean and the standard deviation of $Y = f(X)$:

(5.8)
$$\mu_Y \doteq f(\mu_X)$$
$$\sigma_Y \doteq | f'(\mu_X) | \sigma_X.$$

The approximations (5.8) become the exact relationships (5.5) if f is the linear function $f(x) = a x + b$.

We now give a simple application of the ideas developed so far on transformed variables. Consider a circuit that changes the level of a voltage from X to Y through a function f. I have graphed f in Figure 5.9. There you will also find a graph of a function g that corresponds to a second circuit

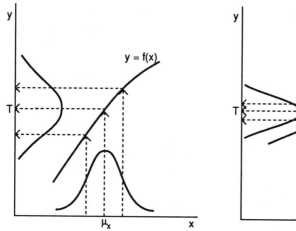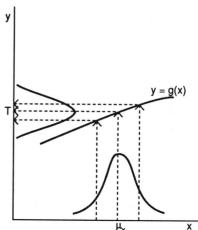

Figure 5.9 Variation Transmission by Two Functions

design. I have sketched a normal distribution on the x-axes as an indication of the variability of the X-characteristic or input. I have also sketched the distributions on the y-axes that result from applying the functions to X.

The line that serves as the linear approximation to f in the neighborhood of μ_X has the equation $y = f(\mu_X) + f'(\mu_X)(x - \mu_X)$. The linear approximation is the equation of line that is tangent to the graph of the function at the point $(\mu_X, f(\mu_X))$. The linear approximation for g is that you get by replacing f by g in the preceding equation of a line. From the linear approximations, $\sigma_Y = |f'(\mu_X)| \sigma_X$ in one case, and $\sigma_Y = |g'(\mu_X)| \sigma_X$ in the other.

Notice that the function g is flatter in the neighborhood of μ_X and transmits less of the variability of X than does f. If the bias of Y from the target voltage T for the circuits is about the same for both f and g, then the circuit g is preferred on the basis of transmitting the smaller variance.

Taguchi (1986, Chapter 6) uses the circuit design called the Wheatstone Bridge to illustrate the concept in a multicomponent model. He stresses engineering design that recognizes the concept of error transmission and the economic benefits that result from having cheaper components (wider tolerance for X) and appropriate functions g to dampen the effect of the increased variance for X.

As a concrete but contrived example, suppose that X has the Binomial distribution B(4,.5) so that $\mu_X = 2$ and $\sigma_X = 1$. Suppose that

$$f(X) = 2X + 1 \text{ and } g(X) = X(4 - X) + 2.$$

If you plot these functions on the interval $0 \le x \le 4$, you will see that g is flatter than f in a neighborhood of x = 2. It follows from (5.5) that

$$\mu_{f(X)} = 5 \text{ and } \sigma_{f(X)} = 2.$$

Exact calculations show that

$$\mu_{g(X)} = 5 \text{ and } \sigma_{g(X)} = \sqrt{1.5}.$$

Hence, less of the variation in X is carried by the function g.

The approximations (5.8) applied to $Y = g(X)$ are not very good here. They result in

$$\mu_{g(X)} \doteq 6 \text{ and } \sigma_{g(X)} \doteq 0.$$

We now reconsider the variability of colorant content in molded plastic bottles. We will use linear approximation in this multivariable situation to further illustrate the concept of the transmission of errors through a function of many variables. The mathematics is somewhat beyond was is expected of the reader so I will essentially give the outline of the problem and a sketch of the results of the analysis.

Example 5.7   Variability in Colorant Content Revisited

Refer to Example 4.11. Let

Y = % red colorant in a bottle
U_1 = weight of red pellets going into the bottle
U_2 = weight of white pellets going into the bottle
U_3 = weight of regrind going into the bottle
V_1 = % of red pellet weight that is colorant
V_2 = % of white pellet weight that is colorant = 0
V_3 = % of regrind weight that is colorant.

Then,

$$Y = \frac{U_1 V_1 + U_2 V_2 + U_3 V_3}{U_1 + U_2 + U_3} \cdot 100.$$

A study provided estimates for the standard deviations of U_2 and U_3. Knowledge of the process history provided an estimate for the standard deviation of V_3. The analysis using the Poisson distribution and reported in Example 4.11 gives a lower bound for the standard deviation of U_1. From specifications for the white and the red pellets, it follows that $V_2 = 0$ and an estimate for the standard deviation of V_3 was determined. Using the multidimensional analog to the standard deviation formula in (5.8), which involves partial derivatives, it was possible to determine that $\sigma_Y \geq 0.18\%$. Since the engineering tolerance for Y is LSL = 2.0% and USL = 2.4%, it is clear that the Y process is not capable. ∎

Section 5.5 References

Feller, William, *An Introduction to Probability Theory and Its Application, Vol. I,* Second Edition, Wiley, New York, 1960.

Ford Motor Co., Continuing Process Control and Process Capability Improvement, Dec., 1987.

Gunter, Berton H., "The Use and Abuse of C_{pk}, Part 1," *Quality Progress,* 72–73, January, 1989a.

Gunter, Berton H., "The Use and Abuse of C_{pk}, Part 2," *Quality Progress,* 108–109, March, 1989b.

Gunter, Berton H., "The Use and Abuse of C_{pk}, Part 3," *Quality Progress,* 79–80, May, 1989c.

Gunter, Berton H., "The Use and Abuse of C_{pk}, Part 4," *Quality Progress,* 86–87, July, 1989d.

Imai, Masaaki, *Kaizen: The Key to Japan's Competitive Success,* Random House Business Division, New York, 1986.

Motorola, "Statistical Process Control," A Document of Motorola, Inc., Motorola Literature Distribution, Phoenix, AZ, 1988.

Rao, C. Radhakrishna, *Statistics and Truth: Putting Chance to Work,* Council of Scientific & Industrial Research, New Delhi, India, 1989.

Taguchi, Genichi, *Introduction to Quality Engineering,* Asian Productivity Organization, Kraus International Publications, White Plains, New York, 1986.

Taguchi, Genichi and Wu, Yuin, *Introduction to Off-Line Quality Control,* Central Japan Quality Control Association, Nagaya, Japan, 1979.

Walton, Mary, *The Deming Management Method,* Dodd, Mead & Company, New York, 1986.

6

Conclusion

I hope that the reading of this book has helped to make it clear that applying statistical concepts in a meaningful way is not a simple matter. The experiences that we have shared have demonstrated that a certain degree of care must be exhibited in performing a statistical study and in interpreting statistical information.

I would hope that the reader now appreciates the importance of pedigrees for data. (I can not recall when I first heard of the term "pedigree" in developing the concept of validity in the philosophy of science; its use in this context is not my invention. The next edition of this book will do the concept more justice and will properly cite the literature in regard to the concept "pedigree".)

Data are essentially useless without their pedigree. If the results of a random sample are reported, ask about the details and the history of the study that resulted in the random sample. Were several random samples not reported in favor of the one that was reported? Was measurement error controlled? What was the population, if the study were of the enumerative type? What was the purpose of the study? Are the persons who conducted the study credible? Do not reach decisions based on reported results and data without first investigating the pedigrees for those results and data.

I hope that the reader now appreciates the advantages to protocol and cooperation in developing statistical information. This is particularly important in decision-making situations where more than one party is affected by the decision that will be based in part on the statistical information.

I wish that more courts and hearing bodies would play the lead in developing the statistical studies of importance to their considerations, if not in specifying protocols. Because of the very nature of a statistical study, particularly, one in which inference is to be made based on the sample information, persons in the adversarial arena have the opportunity to impugn the results for their purposes. I am not concerned about valid criticism and critical analysis. These are essential and part of process control. I am concerned about nit-picking and the confusion that is created when so-called experts present testimony dealing with minutia. Advance agreements in regard to protocols and acceptable challenges make a lot of sense and will ultimately save considerable expense and legal wrangling.

There were positive results based on the action by the Elections Division in the State of Michigan in the 1970's to replace ad hoc procedures for sampling petition signatures with procedures and rules based on sound statistical practice. Professor Jim Stapleton and I designed several plans; the net effect of their implementations and refinements has been positive. The Elections Division can put its efforts into carefully processing a sample of manageable size and can quantify the precision of the projections based on the sample. Attorneys and other interested parties now spend their time concentrating on removing measurement error from a sample that carries the projection rather than in fishing aimlessly about to cast doubts. The level of the arguments has been raised considerably over the years because of the greater focus that the sampling provides. Formerly, persons argued from a base of very uncertain information; formerly, witnesses have paraded before the Board of Canvassers saying that they had checked x signatures and had found y were not registered, having chosen a convenience sample. The Board could make neither head nor tail of this sort of information; no one could.

There is another lesson that we learned from the petition signature experience. Others have reported similar experiences. When a technical person gives an example for purely illustrative purposes, the lay person may choose to make it operational. One example that we gave of a sampling plan and decision rule treated the two errors in the decision process as equally bad. The rule was designed so that the probability of rejecting petitions that were sufficient with an overage of 1% was equal to the probability of certifying petitions with an underage of 1%, both probabilities being .10. We asked the Elections Division staff to consider the errors as to their relative impact in the context of public policy and to consider the size of the error probabilities. I think the staff simply chose a plan that we had given as an example; it happened to treat the errors symmetrically.

One might argue that the error of rejecting petitioners that had presented a sufficient set of signatures is more serious than certifying for the ballot a set that was insufficient. After all, the voters in the general election serve as the final test of the public's sentiment on the issue. On the other hand, there are often very strong proponents and opponents lined-up to argue the merits of a proposal and the Board may wish to take a neutral position with respect to the errors.

I hope that the reader now appreciates the need for operational definitions. Such are essential in statistical studies. Indeed, without clear definitions, confusion and problems will abound in all that we do. Operational definitions serve to limit errors and variation that can render a study useless.

I have not found it possible to include a comprehensive treatment of the implications of the enumerative and analytic distinction in this edition. The idea behind the distinction is pervasive in statistics and its application. At the least, the reader should now appreciate the fact that an analytic study concerns the process; an enumerative study concerns a fixed population of elements that were produced by the process. Information is gained on the process through observation over time and through experiments designed to gather information on the underlying mechanisms and relationships that make-up the process. Information is gained on the fixed population through measurements taken on samples selected from the population.

This distinction is very important. On the technical side, it is related to a useful formula known to every statistician, namely,

$$(6.1) \qquad \sigma_Y^2 = V[E(Y \mid X)] + E[V(Y \mid X)].$$

In words this states that the (unconditional) variance is equal to the variance of the conditional expectation plus the expectation of the conditional variance. The conditional properties correspond to enumerative properties and the unconditional properties correspond to the analytic properties. In Cochran's text on sampling theory, he sometimes gives the conditional properties of estimators based on some outcome and he sometimes gives the unconditional properties of the estimator.

Let Y denote the number of runs in 200 independent tosses of a fair coin. Following Table 4.1, I gave the expected number $E(Y) = 100.5$; the standard deviation is $\sigma_Y = 7.0534$ to four decimal places. This standard deviation was calculated using (6.1). Many books state the properties for the number of runs conditional on the number of H's, which we will denote by X. For the sequence in Table 4.1, $X = 98$. The conditional mean and standard deviation for $Y \mid X = 98$ are 100.96 and 7.0504, respectively.

On the practical side, we realize that risk sharing for errors can be at the process or analytic level when a consumer accepts a shipment based on process characteristics or at the enumerative level when the consumer samples a lot which collects the results of the process.

Suppose a producer produces parts according to a stable process that produces nonconforming parts at a rate of 0.05% and a Consumer agrees to purchase a lot of 10,000 knowing this information about the process. The Consumer puts the lot into production without an enumerative study of the lot and discovers that the lot has 12 nonconforming parts. The Consumer should not be upset with the Producer.

Professor Fabian has made a strong argument in support of judging a physician by the decision-making process that led to the use of a procedure and not by the outcome in the particular case. Of course, information discovered in a surgery and unknowable to the physician may show that, in the particular case, the surgery was not necessary. The physician should not be criticized if he/she used the correct process in arriving at the decision to perform surgery in this case. Here it seems reasonable to judge the physician by the process. (The analytic side to the issue.) Insurance or some other approach should be in place to help the patient or consumer. (The result or enumerative side to the issue.)